Basic Clinical Dialysis

Basic Clinical Dialysis

Edited by

DAVID HARRIS MD (USyd), BS, FRACP
Department of Renal Medicine, Westmead Hospital, NSW

GRAHAME ELDER MBBS (USyd), PhD (UNSW), FRACP
Department of Renal Medicine, Westmead Hospital, NSW

LUKAS KAIRAITIS MBBS (USyd), PhD (USyd), FRACP
Department of Renal Medicine, Westmead Hospital, NSW

GOPALA RANGAN MBBS (UNSW), PhD (USyd), FRACP
Department of Renal Medicine, Westmead Hospital, NSW

The McGraw-Hill Companies

Sydney New York San Francisco Auckland
Bangkok Bogotá Caracas Hong Kong
Kuala Lumpur Lisbon London Madrid
Mexico City Milan New Delhi San Juan
Seoul Singapore Taipei Toronto

McGraw-Hill **Medical**

National Library of Australia Cataloguing-in-Publication data:

Basic clinical dialysis.
Includes index.
For the medical profession.

ISBN 0 074 71501 1.

1. Kidneys—Diseases. 2. Chronic renal failure—Treatment.
3. Hemodialysis. 4. Peritoneal dialysis. I. Harris, David C. H.

617.461059

Published in Australia by
McGraw-Hill Australia Pty Ltd
Level 2, 82 Waterloo Road, North Ryde NSW 2113
Acquisitions Editor: Thu Nguyen
Production Editor: Sybil Kesteven
Editor: Carolyn Pike
Proofreader: Tim Learner
Indexer: Max McMaster
Designer (cover and interior): Jan Schmoeger/Designpoint
Illustrator: Alan Laver, Shelly Communications
Typeset in 9/11 pt Stone Serif by TechBooks, India
Printed on 80 gsm woodfree through World Print Ltd, China

The McGraw·Hill Companies

Contents

Preface

This handbook is intended as a bedside guide to the management of dialysis patients for medical students, resident medical officers, junior advanced trainees in nephrology, nurses and paramedical staff. It is not a comprehensive textbook of dialysis. The book has been written by doctors, nurses, pharmacists and dietitians from the Western Sydney Renal Service, who are based primarily at Westmead, Blacktown and Nepean Hospitals, and by colleagues from Gosford Hospital. The contents summarise protocols that are currently employed in our hospitals. We do not attempt to detail alternative valid approaches to the management of the dialysis patient that may be followed in other units.

List of protocols

List of tables

List of figures

Abbreviations

ACEI	angiotensin-converting enzyme inhibitor
Al	aluminium
ALP	alkaline phosphatase
ANCA	anti-neutrophil cytoplasmic antibody
anti-HBs	hepatitis B surface antibody
anti-HCV	hepatitis C antibody
ANZDATA	Australia and New Zealand Dialysis and Transplant Registry
APD	automated peritoneal dialysis
APTT	activated partial thromboplastin time
ARB	angiotensin receptor blocker
ARDS	adult respiratory distress syndrome
ARF	acute renal failure
ATN	acute tubular necrosis
AV	arteriovenous
β-HCG	beta human chorionic gonadotrophin
BMD	bone mineral density
BMI	body mass index
BMP-7	bone morphogenic protein-7
BP	blood pressure
BSA	body surface area
BSAP	bone-specific alkaline phosphatase
BSL	blood sugar level
BW	body weight
$Ca*PO_4$	calcium by phosphate product
CABG	coronary artery bypass graft
CAD	coronary artery disease

CAPD	continuous ambulatory peritoneal dialysis
CARI	Caring for Australians with Renal Impairment
CCF	congestive cardiac failure
CKD	chronic kidney disease
CK-MB	creatine kinase-MB
CMV	cytomegalovirus
CPAP	continuous positive airways pressure
CRP	C-reactive protein
CT	computerised tomography
cTn-1	cardiac troponin 1
cTn-T	cardiac troponin T
CVC	central venous catheter
CVVHDF	continuous venovenous haemodiafiltration
CXR	chest X-ray
D/P	dialysate/plasma
DDAVP	desmopressin
DFO	desferrioxamine
DTPA	diethylenetriaminepenta-acetic acid
DVT	deep venous thrombosis
DXA	dual-energy X-ray absorptiometry
EBV	Epstein-Barr virus
ECFV	extracellular fluid volume
ECG	electrocardiograph
ESRF	end-stage renal failure
FBC	full blood count
Fe	iron
FFP	fresh frozen plasma
FSH	follicle-stimulating hormone
GBM	glomerular basement membrane
GDP	glucose degradation product
GFR	glomerular filtration rate
GN	glomerulonephritis
GP	glycoprotein
GTT	glucose tolerance test
Hb	haemoglobin
HbA_{1C}	haemoglobin A_{1C}
HbsAg	hepatitis B surface antigen
HBV	hepatitis B virus

HCO_3	bicarbonate
HCV	hepatitis C virus
HD	haemodialysis
HDF	haemodiafiltration
HDL-C	high-density lipoprotein cholesterol
HELLP	haemolysis, elevated liver enzymes, low platelets
HF	haemofiltration
HIT	heparin-induced thrombocytopenia
HITTS	heparin-induced thrombocytopenia plus thrombotic syndrome
HIV	human immunodeficiency virus
HLA	human leucocyte antigen
HP	haemoperfusion
H–PF4	heparin–platelet factor 4
HRT	hormone replacement therapy
HT	hypertension
HUS	haemolytic uraemic syndrome
IBWt	ideal body weight
IM	intramuscular
INR	international normalised ratio
IP	intraperitoneal
IPTH	intact-PTH
ITP	idiopathic thrombocytopenic purpura
IUD	intrauterine device
IV	intravenous
IVC	inferior vena cava
IVI	intravenous injection
JVP	jugular venous pressure
K	potassium
K/DOQI	Kidney Disease Outcomes Quality Initiative
Kt/V	urea clearance
LDL-C	low-density lipoprotein cholesterol
LFT	liver function test
LH	luteinising hormone
LMWH	low-molecular-weight heparin
LTBD	low-turnover bone disease
LV	left ventricular

LVH	left ventricular hypertrophy
Mg	magnesium
MMR	measles, mumps and rubella
MPGN	membranoproliferative glomerulonephritis
MRSA	methicillin-resistant *Staphylococcus aureus*
MRSE	methicillin-resistant *Staphylococcus epidermidis*
MW	molecular weight
Na	sodium
NBM	nil by mouth
NSAID	non-steroidal anti-inflammatory drug
OM	osteomalacia
PD	peritoneal dialysis
PET	peritoneal equilibration test
PPV	pneumococcal polysaccharide vaccine
PSA	prostate specific antigen
PTFE	polytetrafluoroethylene
PTH	parathyroid hormone
QCT	quantitative computerised tomography
RA	right atrium
RBC	red blood cell
RPGN	rapidly progressive glomerulonephritis
RRF	residual renal function
SD	standard deviation
SERMs	selective (o)estrogen receptor modulators
SHBG	sex-hormone binding globulin
SI	sacroiliac
SLE	systemic lupus erythematosus
SVC	superior vena cava
T_3	tri-iodothyronine
T_4	thyroxine
TB	tuberculosis
TG	triglyceride
TSAT	transferrin saturation
TSH	thyroid-stimulating hormone
TTP	thrombotic thrombocytopenic purpura
UF	ultrafiltration
UO	urine output
URR	urea reduction ratio

UTI	urinary tract infection
UV	ultraviolet
VRE	vancomycin-resistant *Enterococcus*
vWF	von Willebrand factor
WBC	white blood cell

Part 1
Introduction

The purpose of this part is to provide a brief summary of normal kidney functions that need to be replaced by dialysis, and to place dialysis within the context of integrated management of the patient with CKD.

Chapter 1.1
Normal kidney function

HIGHLIGHTS

- The normal kidney has complex regulatory, excretory, synthetic and metabolic functions.

The normal kidney helps maintain a stable internal milieu of the body by a range of regulatory, excretory, synthetic and metabolic functions. Dialysis is capable of replacing only some of these functions (see Chapter 1.4). Normal kidney functions include:

1. regulation of:
   (a) ECFV, by excretion or retention of sodium and water
   (b) plasma osmolality, by diuresis or antidiuresis
   (c) acid–base balance, by production and excretion of ammonia, excretion of titratable acids, and reabsorption or excretion of bicarbonate
   (d) potassium balance, especially by its rapid distal tubular secretion
   (e) calcium, phosphate and magnesium balance
2. excretion of:
   (a) solutes of low molecular weight (i.e. <500 Daltons), such as urea and creatinine
   (b) nitrogenous compounds and other ‘uraemic toxins’
   (c) solutes of middle-range molecular weight (i.e. 500–20 000 Daltons)
3. metabolism of filtered proteins, including hormones
4. synthesis of:
   (a) erythropoietin
   (b) renin–angiotensin
   (c) other locally active, haemodynamic hormones, including prostaglandins, nitric oxide and endothelin
5. activation of 25-hydroxycholecalciferol (by 1-hydroxylation)

Chapter 1.2
Stages of chronic kidney disease

HIGHLIGHT

- The stage of CKD predicts clinical manifestations.

Introduction

The damaged kidney can undergo a progressive series of structural and functional adaptations that permit the maintenance of useful function even when damage is severe. The extent and effectiveness of these adaptations characterise the five different stages of CKD. The extent to which dialysis is able to subserve normal kidney function is discussed in Chapter 1.4.

CKD may be defined as kidney damage that has persisted for at least 3 months. It may be staged based on the level of the GFR (Table 1.2A). Clinical evidence of CKD includes abnormal urinalysis and reduced renal size, with or without a reduced GFR.

Table 1.2A Stages of chronic kidney disease

Stage	Clinical definition: kidney damage plus	GFR (mL/min/1.73 m^2)
1	Normal or raised GFR	≥90
2	Mildly reduced GFR	60–89
3	Moderately reduced GFR	30–59
4	Severely reduced GFR	15–29
5	Kidney failure	<15 or dialysis

Staging

Staging is clinically useful as it predicts clinical manifestations, complications and comorbidities of CKD (Table 1.2B), and can guide assessment and management strategies (see Chapter 1.3).

Table 1.2B Clinical manifestations of chronic kidney disease

Stage	Typical clinical manifestations[a]
1	None, or of the primary disease process
2	None; hyperparathyroidism, increased risk of cardiovascular disease
3	Nocturia, anaemia, low 1,25-dihydroxycholecalciferol level, raised serum creatinine level, dyslipidaemia, abnormal ECFV
4	Uraemic symptoms, abnormalities of serum electrolyte levels
5	Severe uraemic symptoms, dialysis

(*a*) These may be seen in any stage beyond the first indicated.

Morbidity and mortality

Morbidity and mortality are worse in patients who start dialysis after late referral to nephrological services compared to those who present earlier in the course of CKD. Referral to a nephrologist should occur by stage 3 of CKD, or earlier when dictated by a need for diagnosis or treatment of the primary disease or the presence of risk factors for the rapid progression of CKD (including diabetes, hypertension, smoking, age $>$65 years, nephrotic range proteinuria).

Many patients can remain at early stages of CKD for a number of years. Hence, the majority of patients with CKD die, mainly from cardiovascular disease, prior to the need for dialysis.

Reference

1. National Kidney Foundation K/DOQI. Clinical practice guidelines for chronic kidney evaluation, classification and stratification. Am J Kidney Dis 2002; 39 (suppl 1): S17–S31.

Chapter 1.3
Assessment of the patient with chronic kidney disease

HIGHLIGHTS

- Patients with CKD require comprehensive assessment.
- Assessment and management is guided by the stage of CKD.

Assessment and management

There should be a smooth continuum in the management of patients with CKD throughout its stages and during dialysis. Patient assessment and management can be guided by the stage of CKD (Protocol 1.3A).

PROTOCOL 1.3A

How to assess and manage patients according to the stage of chronic kidney disease

Stage of CKD	Based on GFR[(a)]	Direct assessment and management to[(b)]
1	≥90	primary disease, cardiovascular risk
2	60–89	early hyperparathyroidism, progression of CKD
3	30–59	anaemia, dyslipidaemia, ECFV
4	15–29	electrolyte abnormalities, preparation for dialysis and transplantation
5	<15	complications of advanced CKD and dialysis

(*a*) In mL/min/1.73 m^2.
(*b*) May apply for any stage beyond that in which first mentioned.

Initial assessment

Initial assessment of the patient with CKD must include a complete history and examination, basic blood and urine tests, renal structural imaging and other tests as dictated by the cause and stage of CKD, and the results of the initial assessment (see Protocol 1.3B).

PROTOCOL 1.3B

How to initially assess the patient with chronic kidney disease

- Full personal and family medical history.
- Comprehensive physical examination.
- Serum biochemistry and full blood count.
- Urinalysis (for protein, glucose, blood, leucocytes, nitrite), urinary sediment, urine protein:creatinine or albumin:creatinine ratio.
- Renal ultrasound.
- Other tests based on cause and stage of CKD (see Protocol 1.3A).

Comprehensive assessment

The main principles of a comprehensive assessment of the patient with CKD are summarised in Protocol 1.3C.

Establish the cause

Establish the cause of CKD as many diagnoses carry additional implications, including a familial nature and recognised complications (see below).

Differentiate from acute kidney disease

Differentiate CKD from acute kidney disease by means of renal ultrasound, haemoglobin level and serial assessment of renal function. The presence of a small renal size, a loss of corticomedullary differentiation and an increased renal

PROTOCOL 1.3C

How to comprehensively assess the patient with chronic kidney disease

- Establish the cause of CKD.
- Differentiate from acute kidney disease.
- Quantify GFR.
- Calculate the rate of progression of CKD.
- Quantify urinary protein excretion.
- Assess cardiovascular risk.
- Look for reversible renal dysfunction.
- Assess lifestyle risks.
- Look for specific complications of the primary disease.
- Assess suitability for dialysis.
- Assess suitability for transplantation.
- Assess medications.

echogenicity on ultrasound, normochromic normocytic anaemia and hyperphosphataemia, and a reduction in GFR for more than 3 months are indicative of chronic disease.

Examine urinary sediment

Examine the urinary sediment in a fresh centrifuged sample, which has been transported in boric acid to preserve casts. The presence of red or white cell casts indicates an inflammatory process, usually acute, while broad casts are suggestive of advanced renal disease.

Quantify GFR

Quantify the GFR to assign the stage of CKD. This is usually done by using the Cockcroft–Gault formula to first determine the uncorrected creatinine clearance:

$$\text{Creatinine clearance (males)(mL/min)} = \frac{(140 - \text{age}) \times \text{body weight (kg)}}{0.814 \times \text{plasma creatinine } (\mu\text{mol/L})}$$

For females, this equation is multiplied by 0.85.

The creatinine clearance is then multiplied by 1.73 m^2 (the standard body surface area, BSA) and divided by the patient's BSA to give the GFR in mL/min/1.73 m^2. (The term GFR can be used interchangeably with creatinine clearance for convenience.)

Alternatively, the GFR may be estimated from the creatinine clearance using a 24-hour urine collection:

$$\text{Creatinine clearance (mL/min)} = \frac{\text{urinary creatinine } (\mu\text{mol/L}) \times \text{urine volume (mL)}}{\text{plasma creatinine } (\mu\text{mol/L}) \times \text{time of collection (min)}}$$

The adequacy of the collection can be estimated from the creatinine index (i.e. the expected creatinine excretion), which should be 180–230 μmol/kg/d for males and 120–190 μmol/kg/d for females. Measurement of the GFR using isotopic methods (e.g. technetium DTPA clearance) is more accurate and should be considered at least once to check the accuracy of the above creatinine-based methods.

Calculate the rate of progression

Calculate the rate of progression of CKD by serial (e.g. quarterly) calculation of the GFR. Look for factors that may accelerate its progression and also at how adequate the treatment is at slowing the progression (e.g. glucose control, blood pressure control, minimisation of proteinuria).

Quantify urinary protein excretion

Quantify the urinary protein excretion by an initial timed urinary collection for protein and creatinine, and follow the response to anti-proteinuric therapy with assessment of serial (e.g. quarterly) spot urinary protein:creatinine ratios or albumin:creatinine ratios.

Assess cardiovascular risk

Assess cardiovascular risk by personal and family history and examination relevant to the heart and vasculature (including blood pressure and smoking history). Measure plasma lipid levels, perform ECG and echocardiography for evidence of left ventricular hypertrophy, and Doppler assessment of the carotid, abdominal and lower limb vessels as suggested by clinical assessment, and possibly measure non-traditional risk factors (e.g. homocysteine, lipoprotein (a), CRP, serum calcium by phosphate product) (see Chapter 2.1). Renal disease is a major risk factor for cardiovascular disease, and cardiovascular disease is the major cause of morbidity and mortality in patients with CKD.

Look for reversible renal dysfunction

Look for factors causing acute, reversible deterioration of renal function, including abnormal ECFV (usually volume depletion), hypotension or severe hypertension, cardiac failure, lower urinary tract obstruction (e.g. stones, sloughed papillae, malignancy), systemic sepsis, electrolyte derangements (e.g. hypercalcaemia), nephrotoxic drugs (e.g. ACEIs, ARBs, NSAIDs, coxibs, antibiotics) and other nephrotoxins (e.g. radiocontrast media).

Assess lifestyle risks

Assess lifestyle factors that might contribute to comorbidity, including body habitus (BMI), diet, smoking and exercise.

Complications of the primary disease

Consider the possibility of complications relevant to the primary disease, including:

- diabetes—macrovascular and microvascular disease
- primary and secondary glomerulonephritides—acute disease flare
- polycystic kidney disease—cerebral aneurysm (new or atypical headache, family history), obstruction (calculi), infection (UTI, cyst) and cyst (rupture, infection, bleed)
- reflux nephropathy—UTI
- renovascular disease—renal artery occlusion
- analgesic nephropathy—obstruction (sloughed papilla), transitional cell carcinoma

Complications relevant to the stage of CKD

Look for complications relevant to the stage of CKD, including:

Stage 2	abnormal calcium, phosphate and PTH levels (see Chapter 2.4); hypertension
Stage 3	low 25-hydroxy and 1,25-dihydroxycholecalciferol levels (see Chapter 2.4); anaemia (see Chapter 2.3); fluid overload
Stage 4	abnormal electrolytes—potassium, bicarbonate, uric acid, magnesium
Stage 5	clinical evidence of bleeding diathesis (see Chapter 2.3), serositis, sexual dysfunction, neuropathy, malnutrition (see Chapter 2.10)

These assessments should be made serially (e.g. quarterly) from the stage at which the abnormality is expected to arise for the first time.

Assess suitability for dialysis

From stage 4, assess suitability for dialysis, including:

- sites for dialysis access (see Chapters 3.4, 4.3)
- personal coping mechanisms, social supports, transport and flexibility of employment
- hepatitis B and C, HIV, MRSA and VRE status, and vaccination status (see Chapter 2.6)

Assess suitability for transplantation

From stage 4, assess suitability for transplantation (see Chapter 2.8), including:

- potential living donors
- risk of malignancy
- cardiovascular risk
- other significant comorbidity

Assess medications

Regularly assess the appropriateness of all medications and commence a personal medication list for each patient.

Reference

1. National Kidney Foundation K/DOQI. Clinical practice guidelines for chronic kidney evaluation, classification and stratification. Am J Kidney Dis 2002; 39 (suppl 1): S17–S31.

Chapter 1.4
Limitations of dialysis as replacement for normal kidney function

HIGHLIGHT

- Dialysis requires supplementary treatment to replace most kidney functions.

The ability of dialysis to replace normal kidney functions depends, in part, on residual renal function. No normal kidney function is completely reproduced or replaced by dialysis, and most patients require additional therapy (Table 1.4). Suboptimal patient compliance with fluid intake, diet and medications is the major cause of apparent failure of dialysis to replace normal kidney functions.

Dialysis procedures are associated with complications (see Chapters 3.7, 4.8, 4.9), which further limit the ability of dialysis to replace normal kidney function (e.g. loss of water-soluble vitamins).

Table 1.4 Supplementary therapy required with dialysis

Function	Supplementary therapy
Regulation of:	
• ECFV	Appropriate sodium and water intake, loop diuretic (if passing urine)[(a)]
• plasma osmolality	Appropriate water intake[(a)]
• acid–base balance	None (occasionally $NaHCO_3$ supplementation with HD)
• potassium balance	Dietary restriction (occasionally Resonium A) with HD,[(b)] sometimes dietary supplementation with PD
• calcium/phosphate balance	Phosphate binders,[(b)] calcitriol
• magnesium balance	Avoid excessive magnesium
Excretion of small solutes	None
Excretion of larger solutes	None available[(c)]
Metabolism of proteins	None available and none required
Synthesis of:	
• erythropoietin	Erythropoietin or analogues
• renin–angiotensin	Sometimes ACEI or ARB
• local hormones	None available and none required
Activation of 25-hydroxy cholecalciferol	Calcitriol

(*a*) Based on assessment of ideal body weight and urine output.
(*b*) Not required with daily HD (phosphate supplementation may be required).
(*c*) High-flux HD and HDF are superior to conventional HD and PD in removing larger solutes.

Chapter 1.5
Preparation of the patient for maintenance dialysis

HIGHLIGHTS

- Optimal patient preparation for dialysis requires predialysis education (in group and individual sessions), individualised assessment for dialysis modality, assessment and reduction for infection risk, and timely placement of dialysis access (Protocol 1.5).
- If possible, dialysis should be performed at home.

PROTOCOL 1.5

How to optimally prepare a patient for maintenance dialysis

- Provide timely group and individualised predialysis education.
- Assess social and psychological well-being.
- Assess suitability for home dialysis.
- Assess suitability for PD versus HD.
- Correct reversible factors precluding chosen dialysis modality.
- Assess and treat risk of infectious diseases (hepatitis B and C, HIV, MRSA, VRE).
- Create dialysis access to allow optimal time of initiation.
- Obtain consent for ANZDATA collection.

Education

Aims

The aims of predialysis education are to:

- provide information to enable patients to make an informed choice as to which dialysis modality would suit their lifestyle best
- promote home dialysis as the first option
- promote a philosophy of independence and responsibility for personal health

Structured education should commence when the GFR is 20 mL/min or is deteriorating steadily.

Education can be provided using:

- written material
- Internet material
- group sessions with other predialysis patients and family members or close friends
- individual sessions with a dialysis nurse specialist

Group sessions

Group sessions should:

- provide an opportunity for patients to meet others with similar health problems and to share experiences
- include as speakers nurses, pharmacists, dietitians, patients and partners who perform dialysis at home on PD or HD
- allow discussion of the advantages and disadvantages of different dialysis modalities, including issues that might make a particular form of treatment very difficult or impossible, such as abdominal surgery with peritoneal membrane scarring or adhesions, frequent overseas travel for work, or occupation (e.g. swimming instructor)
- include existing dialysis patients to discuss issues of management of life on dialysis, such as work, leisure, family and children, travel and holidays

Individual sessions

Individual sessions should:

- follow the group session
- answer questions related to the information given in the group session
- check the patient's understanding of dialysis treatment options
- include a visit to PD and HD training areas, and discussion with patients in training

Choice of dialysis modality

Assessment

Nursing assessment is important to enable the appropriate choice of dialysis modality. Assessment should identify the likely contraindications to home dialysis therapy, such as:

- mental confusion or dementia or untreatable psychiatric disorder
- visual impairment (unless assistance is available)
- impaired manual dexterity
- poor living conditions or homelessness

Assessment should identify other likely contraindications or problems related to PD or home HD (Table 1.5). See also Chapters 3.1 and 4.1 for comparisons between PD and HD.

Individual assessment is important as it may allow timely identification of problems and relative contraindications that can be reversed.

Home dialysis

Where family involvement is required for home dialysis therapy, it is essential that the family understands that long-term commitment is involved. Home dialysis is the preferred option when there are no contraindications, as it:

- maintains patient independence

Table 1.5 Relative and absolute contraindications to home therapy

Peritoneal dialysis	Home haemodialysis
Inadequate peritoneal membrane surface area	Vasculature unsuitable for AV fistula
Previous abdominal surgery with adhesions	Needle phobia
Unrepaired hernia	Medical comorbidities, in particular cardiovascular
Pleuro-peritoneal communication	Non-English speaking unless English-speaking assistant is available (ability to communicate by telephone is essential)
Bowel problems (e.g. chronic constipation, diverticulitis)	Inadequate or poor quality water supply
Severe respiratory insufficiency	Accommodation—hostel or nursing home
Ileal conduit or colostomy	Accommodation—inadequate space for the machine
Abdominal obesity	Occupation—involves travel
Large muscle mass	
Minimal or no residual renal function	
Body image issues related to catheter	
Lifestyle—water sports	

- empowers patients, with respect to their treatment regimen, to lessen the impact on their lifestyle and to allow more dialysis hours (e.g. second daily, daily, overnight) than are usually available with satellite or in-centre dialysis
- reduces the requirement for dialysis staff

Satellite program

If home dialysis therapy is impossible, patients are treated through an in-centre or satellite program. Issues such as transport and time commitments (e.g. to children, study and work) need to be addressed.

Dialysis not an option

For some patients long-term dialysis is a poor option, either by choice or because renal failure per se is not the major component of the patient's health problems. Long-term dialysis should enable a patient to gain quality from life, rather than just time without quality (see Chapter 1.6).

Timely creation of dialysis access

Once the preferred dialysis modality has been chosen, the patient should be referred to a surgeon for assessment and planning of vascular access surgery or Tenckhoff catheter insertion.

Initiation

Optimal initiation of dialysis involves a smooth transfer from predialysis care on to a maintenance dialysis program, and avoidance of unplanned dialysis in an acute care facility.

Before commencement of dialysis:

- determine hepatitis B and C, HIV and VRE status (see Chapter 2.6)
- determine vaccination status (see Chapter 2.6)
- involve social worker and/or psychologist if indicated
- identify and repair hernias in potential PD patients
- treat and prevent constipation in potential PD patients

Indications for initiating dialysis

Indications for initiating dialysis include:

- GFR of 6–12 mL/min, depending on symptoms
- symptoms, including lethargy, anorexia, nausea, vomiting or foul taste
- loss of muscle mass or evidence of malnutrition
- difficult to control hypertension and fluid overload
- refractory metabolic complications

Ideally, dialysis should be started before the patient is symptomatic. However, a balance is required between starting dialysis too early with the risk that the patient may

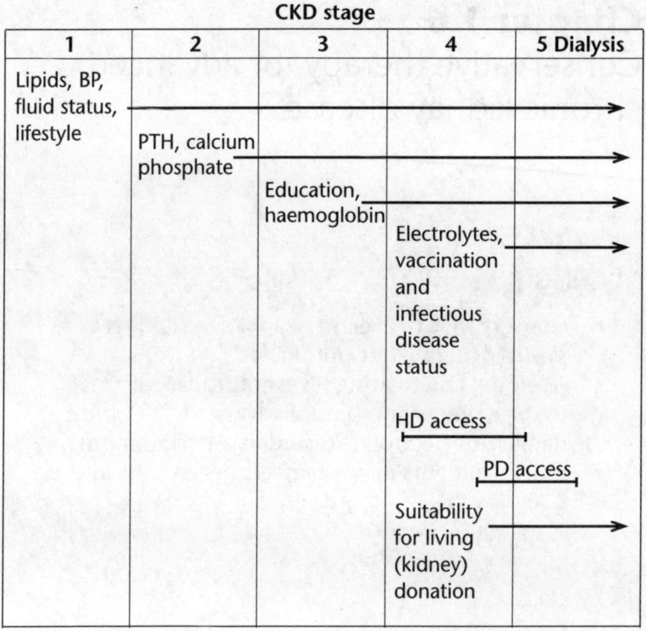

Fig. 1.5 Checklist for management of CKD and patient preparation for dialysis

fail to notice any benefit from dialysis, and starting too late with the risk of adverse consequences to long-term cardiac and bone health.

Poor planning for long-term dialysis treatment and delays in initiation may contribute to the risk of the patient remaining with in-centre or satellite programs, and to an increased risk of morbidity and mortality.

References

1. Baxter's kidney directions. Info renal. Online. Available: <http://www.kidneydirections.com>.
2. Written patient information, produced by Baxter Healthcare, Fresenius Medical Care and Gambro Pty Ltd (available in the dialysis unit).

Chapter 1.6
Conservative therapy for advanced chronic kidney disease

HIGHLIGHTS

- Conservative treatment of advanced CKD is a positive therapeutic option.
- Avoiding and treating reversible causes of renal dysfunction, symptoms of advanced CKD and dangerous electrolyte disorders are paramount.
- Elderly patients may, against expectations, tolerate dialysis well.

Definition

'Conservative' treatment of advanced CKD is a positive therapeutic option for patients in whom dialysis is unlikely to prolong or improve quality of life. Advanced age and comorbidity are the usual reasons for opting for conservative advanced CKD treatment. However:

- in an elderly patient, good compliance may partially compensate for the adverse effects of dialysis
- a patient may inappropriately opt for conservative therapy because of unfounded fears about dialysis or depression
- it is very difficult to predict with any accuracy which patients who have been assessed as borderline candidates for dialysis will or will not achieve improved quality of life or 'useful' additional longevity with dialysis

Uses of conservative therapy

Conservative treatment of advanced CKD may be surprisingly effective at prolonging useful and enjoyable life. Conservative treatment should be directed not only at symptom control and avoiding danger, but also at delaying further progression of CKD, its complications and comorbidities.

Principles of conservative therapy

The principles of conservative CKD therapy are to prolong and improve quality of life. These include:

- avoidance of all nephrotoxins (e.g. drugs, radiocontrast media)
- regular assessment of all medications to avoid polypharmacy, nephrotoxicity (e.g. NSAIDs, coxibs, ACEIs, ARBs, excess vitamin D), inappropriate dosing of renally excreted drugs (e.g. allopurinol, digoxin) and other contraindicated medications (e.g. metformin—risk of lactic acidosis)
- avoidance or rapid reversal of acute precipitants of renal dysfunction, including abnormalities of ECFV and BP
- non-dialytic treatment for the symptoms of CKD:
  - dietary protein restriction for anorexia, nausea and vomiting
  - correction of the symptomatic abnormality of ECFV, especially loop diuretics for fluid overload
  - erythropoietin or its analogues for anaemia
  - antidepressants
  - carbamazepine for dysaesthesiae, gabapentin for restless legs, chlorpromazine for hiccoughs, clonazepam for uraemic myoclonus
  - phosphate binders, antihistamines and UV therapy for pruritus
- correction of dangerous or symptomatic electrolyte abnormalities:
  - dietary restriction and Resonium A for hyperkalaemia

  - phosphate binders to control hyperphosphataemia (risk of pruritus, metastatic calcification), appropriate use of calcium and calcitriol for hypocalcaemia (risk of negative inotropy and arrhythmias) and to reduce the risk of worsening bone disease (see Chapter 2.4).
  - oral sodium bicarbonate to treat severe acidosis (risk of worsening bone disease and malnutrition)
- possible continuation of therapy aimed at slowing the progression of CKD, including ACEIs and ARBs, as these may be effective, even in patients with advanced CKD; however, hyperkalaemia and an acute haemodynamic reduction of the GFR may preclude their use; advise patients to cease them temporarily when at risk of reduced ECFV (e.g. vomiting, diarrhoea, sepsis)
- adequate pain relief
- regular multidisciplinary assessment of social functioning

Palliative care

Death from end-stage CKD usually involves progressive obtundation. As long as severe symptoms (especially of fluid overload) are avoided, the patient should not suffer. Opioid analgesics should be used as required.

PROTOCOL 1.6

How to optimise therapy for advanced chronic kidney disease without dialysis

- Be certain that dialysis is not worthwhile.
- Avoid and treat reversible causes of renal dysfunction.
- Treat symptoms.
- Treat dangerous electrolyte abnormalities.
- Consider continuing treatment to slow CKD progression.
- Assess and optimise social functioning.
- Consider the need for a social worker, geriatrician and palliative care team.
- Plan for a calm, peaceful and dignified death, with minimal suffering.

Part 2
General care of patients with chronic kidney disease

Dialysis overcomes the immediate, lethal consequences of renal failure and maintains life but falls far short of 'renal replacement therapy'. This part outlines the hormonal, bone and mineral, vascular and nutritional complications of chronic kidney disease and provides clinical algorithms for their management. Early recognition and management of the less immediate (and easily overlooked) consequences of CKD may reduce serious complications. Regular discussion between nursing, dialysis, dietetic, clinical pharmacology and medical staff and routine surveillance procedures are likely to provide the best patient outcomes. This part also discusses the management of pregnant patients and guidelines for assessing patients for transplantation, which provides improved prospects for quality of life and long-term survival.

Chapter 2.1
Cardiovascular risk management

HIGHLIGHTS

- The mean age of commencement of dialysis in Australia in 2002 was 59 years, with 55% of patients over 60 years of age.
- Coronary artery disease was proven or suspected in 40% of these patients and 37% had diabetes.
- Nearly half of the deaths in dialysis and post-transplantation patients are due to cardiac causes. Even when adjusted for age, sex and diabetes, cardiovascular mortality in patients on dialysis is 10–20-fold that of the general population (Fig. 2.1).

Aetiology of cardiovascular disease

Table 2.1 lists the risk factors that influence the development of cardiovascular disease, including those exacerbated by CKD.

Presentation

- Typical presentation is angina.
- Recurrent dialysis-related hypotension may be a sign of coronary disease.
- Breathlessness occurs due to pulmonary oedema.
- Arrhythmias.
- Sudden death due to silent coronary disease and arrhythmias.

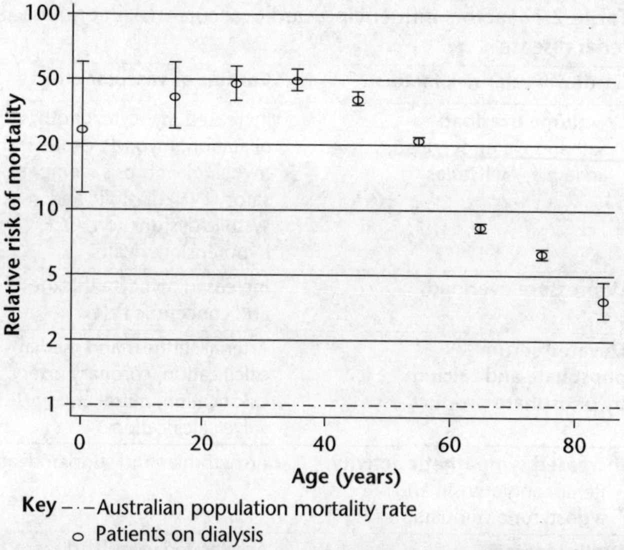

Fig. 2.1 Age-specific mortality rates for patients treated with dialysis relative to the Australian population rates, for the 5-year period 1997–2002. Source: McDonald S. Chapter 3. Deaths. 26th Annual Report. 2003 Report. Adelaide: ANZDATA, 2004.

Investigations

Protocol 2.1A lists the investigations to perform if CAD is suspected.

Management goals

Protocols 2.1B and 2.1C show how to minimise cardiovascular risk factors.

Table 2.1 Factors influencing cardiovascular risk in end-stage renal disease

Cardiovascular risk factors	Cardiac or vascular effect
LV volume overload Salt and water overload, anaemia, AV fistulas	Increased myocyte length, LV dilatation, myocyte death and myocardial fibrosis, compensatory LVH, diastolic and systolic dysfunction, CCF, hypotension, death
LV pressure overload	Increased myocyte thickness and concentric LVH
Elevated serum phosphate and calcium by phosphate product	Arterial stiffness and medial calcification, coronary artery calcification, mitral and aortic valve calcification
Increased sympathetic activity Renin–angiotensin and aldosterone stimulation	Arrhythmias and sudden death
Dyslipidaemia HD: 20–40% of patients elevated TG and reduced HDL-C levels PD: 25–50% of patients elevated cholesterol, LDL-C, TG, reduced HDL-C levels PD lipid profile more atherogenic than HD profile	Accelerated vascular disease
Smoking, obesity, diabetes, sedentary lifestyle	Accelerated vascular disease
Potential risk factors Elevated levels of lipoprotein (a) (more often raised in patients on PD than HD) Elevated levels of homocysteine Low levels of fetuin-A (a calcium-binding protein that reduces calcium precipitation) Low levels of BMP-7 (an inhibitor of the transdifferentiation of vascular smooth muscle cells to cells with an osteoblast phenotype)	Unproven effects

PROTOCOL 2.1A

How to evaluate patients on dialysis for suspected coronary artery disease

- CAD should be excluded in patients developing unexpected fluid overload, pulmonary oedema and hypotension, as well as in all patients awaiting transplantation.
- Perform CXR, ECG and collect blood for cardiac troponin T[(a)] and CK-MB.
- Perform cardiac echography at dry weight (preferably the day after dialysis) to assess for wall abnormalities and systolic/diastolic function.
- Perform dipyridamole (Persantin) sestamibi scan.
- Carry out coronary angiography in patients with abnormal dipyridamole sestamibi scan results who are fit enough to undergo revascularisation.

(*a*) Cardiac troponin T (cTn-T) is reported to be elevated in up to half of all patients with renal disease and cTn-1 in up to one-quarter. Factors influencing cTn-T levels include reduced clearance in renal failure, increased myocyte death due to cardiac failure and increased LVH. Elevated levels identify patients at greater risk of all-cause mortality while normal levels of troponin are a good prognostic factor. cTn-T levels tend to increase after dialysis. Repeat testing to detect a change in the cTn-T level is helpful for interpretation.

PROTOCOL 2.1B

How to minimise cardiovascular risk factors in patients undergoing dialysis

- Achieve and maintain dry weight.
- Aim for target BP levels ≤130/85 mmHg for patients on CAPD or post-HD (ambulatory BP correlates best with post-HD BP).
- For patients known to have cardiovascular disease, maintain Hb levels in range 110–120 g/L with an erythropoietic agent if necessary (Chapter 2.2).
- Aim for target values of phosphate, calcium, calcium by phosphate product and PTH (Chapter 2.4).
- Aim for reduction of traditional risk factors: smoking, sedentary lifestyle and obesity, diabetes and dyslipidaemia.
- For patients on dialysis with dilated cardiomyopathy, carvedilol has been reported to improve cardiovascular function and mortality over 24 months.

PROTOCOL 2.1C

How to minimise cardiovascular risk factors: application of targets not derived from chronic kidney disease data

- Aim for HbA_{1C} of <7–7.5%.
- Aim for total cholesterol level of ≤4 mmol/L, LDL-C ≤2.5 mmol/L, TG <2 mmol/L and HDL-C >1 mmol/L with diet, statins or fibrates.[(a)]
- Statin therapy is recommended for all patients with coronary heart disease. Statins have been shown to be beneficial and generally well tolerated.
- Aim for vitamin B_{12} and RBC folate levels in the normal range to reduce homocysteine levels.
- For patients with known ischaemic heart disease, low-dose aspirin should be used unless contraindicated.
- For patients with systolic heart failure, ACEI and low-dose beta-blockers should be used unless contraindicated. ARBs can be used in combination with or as replacement for an ACEI.
- For patients with diastolic heart failure, carvedilol has been reported to improve diastolic function and candesartan may reduce admission for heart failure.[(b)]

(*a*) If the effect of treatment in patients with CKD were similar to the effect in the general population, a higher absolute risk reduction would be anticipated due to the higher baseline risk. However, the '4D Trial' was recently reported to show no treatment benefit of atorvastatin versus placebo in patients with type 2 diabetes on dialysis.
(*b*) SWEDIC and CHARM-preserved trials.

Hypertension management

About 90% of patients undergoing dialysis have HT. At the start of dialysis the prevalence of HT is higher in those patients on HD than on PD. However, long-term prevalence rates are similar, probably due to the gradual loss of residual function while on PD.

For general management guidelines of HT see Protocols 2.1D and 2.1E.

PROTOCOL 2.1D

How to manage hypertension in patients undergoing dialysis

- Careful assessment of the 'dry' or 'ideal' body weight (see Protocol 3.3E).
- Gradual reduction of this target weight by 0.5–1 kg weekly.
- Fluid removal on dialysis should be slow and consideration should be given to more dialysis hours. Overnight dialysis often leads to drug-free BP control.
- Salt intake should be reduced.
- The interdialytic weight gain should be limited to <2 kg if possible.
- Avoid increased doses of epoetin unless the BP is <160/100 mmHg. Consider withholding epoetin therapy.
- Alpha-blockers and calcium channel blockers are effective and generally do not result in hypotension.
- Beta-blockers, which block sympathetic activity, may cause hypotension during HD.
- ACEIs and ARBs may be less effective in elderly patients and when patients have fluid overload and suppression of the renin–angiotensin system. However, these drugs may reduce the rate of loss of residual renal function.

PROTOCOL 2.1E

How to manage severe hypertension

- Severe HT should generally be managed in hospital.
- Unless the BP is >180/120 mmHg, acute medical treatment is not usually required, except when symptoms such as headache or visual disturbance are present.
- In a patient with fluid overload, remove fluid to the IBWt. This target weight may need to be redefined.
- The BP generally responds to oral antihypertensive agents. Aim for a smooth reduction in BP. Beware of high doses of antihypertensive drugs that can result in hypotension and stroke.
- If the patient is already at his or her IBWt, severe HT can be treated with IV hydralazine (see below) or clonidine.
- Treat with hydralazine 5 mg IVI (onset of action 5–10 minutes). The dose may be repeated every 15–20 minutes or doubled if there is no response. Notify the consultant if this treatment is used.

Post-dialysis hypotension

Management of dialysis-associated hypotension is covered in Protocol 3.7A. If hypotension occurs after dialysis, the IBWt should be reassessed and antihypertensive drugs reduced. Ceasing beta-blockers may be useful because these drugs block baroreceptor responses and sympathetic activity. On the other hand, calcium channel blockers and alpha-blockers generally do not worsen hypotension. Many antihypertensive drugs are effective for 24 hours, so omitting doses on the morning of dialysis may not effectively reduce hypotensive episodes. Patients with recurrent hypotension should have their cardiac function reassessed.

Angina

Angina is usually due to critical CAD and may be exacerbated by anaemia.

In a quarter of HD patients, angina is due to non-atherosclerotic disease, such as:

- cardiomyopathy
- small vessel disease (HT, diabetes, vascular calcification)
- LVH with reduced coronary reserve

Management of angina is along standard care guidelines:

- beta-blockers, calcium channel blockers, low-dose aspirin (usefulness uncertain)
- long-acting nitrates (may be limited by hypotension during dialysis)
- correct anaemia (target Hb level 110–120 g/L)
- PD may be preferable to HD

Coronary artery surgery/angioplasty in dialysis patients

- Dialysis patients have a higher rate of complications after surgery and restenosis after angioplasty.
- CABG is generally preferred over angioplasty ± stenting, particularly in patients with diabetes.
- CABG results in better long-term survival but more perioperative complications.
- Contrast media used during angioplasty and postoperative hypotension may cause loss of residual renal function. This may be particularly important in PD patients.
- Drug eluting stents are promising, with fewer clinical events due to restenosis.

References

1. De Zoysa JR. Cardiac troponins and renal disease. Nephrology 2004: 9(2): 83–8.
2. Lipid management guidelines—2001. Med J Aust 2001; 175(5): S72.
3. The Heart Foundation. Online. Available: <http://www.heartfoundation.com.au>.
4. ANZDATA Registry. Online. Available: <http://www.anzdata.org.au>.

5. Bergstrom A, Andersson B, Edner M, et al. Effect of carvedilol on diastolic function in patients with diastolic heart failure and preserved systolic function. Results of the Swedish Doppler-echocardiographic study (SWEDIC). Eur J Heart Fail 2004; 6(4): 453–61.
6. Yusuf S, Pfeffer MA, Swedberg K, et al. Effects of candesartan in patients with chronic heart failure and preserved left-ventricular ejection fraction: the CHARM-Preserved Trial. Lancet 2003; 362(9386): 777–81
7. Cice G, Ferrara L, D'Andrea A, et al. Carvedilol increases two-year survival in dialysis patients with dilated cardiomyopathy: a prospective, placebo-controlled trial. J Am Coll Cardiol 2003; 41(9):1438–44.
8. Wanner C. Results of a randomized controlled trial with atorvastatin in dialyzed diabetic patients (4D Trial). Renal Week 2004. Symposium: Atherosclerosis: A different disease in CKD than in the general population. Am Soc Nephrol 29 October 2004.

Chapter 2.2
Erythropoiesis

HIGHLIGHTS

- Most patients on dialysis have anaemia unless they are treated with an erythropoietic agent. These patients have a reduced erythropoietin response to lowering of the Hb level and inadequate erythropoiesis. In addition, the rate of RBC destruction is increased.
- Iron deficiency commonly contributes to the anaemia seen in patients with CKD and influences the response to erythropoietic agents. Oral iron therapy fails to replete iron stores in many dialysis patients.
- Other factors include inadequate dialysis, secondary hyperparathyroidism, chronic inflammation and pro-inflammatory cytokine release, which may result in a 'functional iron deficiency' with iron stores unable to be fully utilised.
- For patients without cardiovascular disease, CARI guidelines suggest a minimum Hb concentration of 110 g/L and a preferred Hb target range of 120–140 g/L.
- The preferred Hb target range for patients on dialysis with cardiovascular disease is 110–120 g/L.
- Ferritin levels should be maintained above 100 μg/L. However, higher serum ferritin values of 300–800 μg/L and transferrin saturations (TSATs) of 20–50% may be beneficial for optimising the response to epoetin.

Characteristics of uraemic anaemia

- Anaemia is often present at GFRs below 60 mL/min. Anaemia is very common at GFRs below 30 mL/min.
- Plasma erythropoietin levels are inappropriately low for the Hb level.
- For a given level of renal dysfunction, patients with diabetic nephropathy or tubulointerstitial disease may manifest a greater degree of anaemia, whereas patients with polycystic kidney disease often maintain a normal or near normal Hb concentration.
- Males tend to have higher Hb concentrations than females due to the erythropoietic effect of androgens and due to menstrual blood loss in premenopausal females.

Benefits of correcting uraemic anaemia

- Improved quality of life.
- Increased exercise capacity.
- Shortened skin bleeding time and improved platelet function.
- Reduction in LVH.
- Improvement in libido.
- Improvement in appetite.
- Reduction in blood transfusions.

Haematinic deficiency

It is important to exclude haematinic deficiency. Iron deficiency, the most common haematinic deficiency, may be caused by:

- poor dietary intake
- chronic gastrointestinal blood loss exacerbated by antiplatelet therapy and heparin
- iron chelation by oral phosphate binders
- recurrent blood testing
- repetitive blood loss on HD

Other causes of haematinic deficiency are folic acid and vitamin B_{12} deficiency. Folic acid is a water-soluble vitamin and is removed by dialysis. Vitamin B_{12} has a molecular weight of 1355 Daltons and is also removed by dialysis.

Assessment of iron status

Ferritin levels reflect the adequacy of iron stores but, because ferritin is an acute phase reactant, ferritin levels require interpretation. Serum iron testing is unreliable because of diurnal variation.

To exclude iron deficiency the following should be present:

- ferritin >100 μg/L
- TSAT (a measure of circulating iron) >20%
- percentage of hypochromic RBCs, a marker of iron availability, should be <10% (this test is not widely used)

However, for optimal responses to epoetin therapy, higher target values than those used to exclude iron deficiency may be useful:

- serum ferritin level 300–800 μg/L
- TSAT 20–50%
- percentage of hypochromic RBCs <2.5%

PROTOCOL 2.2A

How to give intravenous iron

- In patients with iron deficiency (ferritin level <100 μg/L or TSAT <20%), give iron polymaltose.
- For HD patients, give 100 mg/week (over 5 minutes) for 5 weeks.
- For PD patients, iron polymaltose can be given as a slow 500–1000 mg infusion over 4–6 hours.
- Adverse effects may include flushing and bronchospasm.
- Patients with results for iron studies in the target range generally require around 100 mg iron polymaltose every 1–3 months (or less frequently as a higher dose infusion).
- For patients with iron overload (ferritin level >800 μg/L or TSAT >50%) cease IV iron for 3 months.
- Iron sucrose can be administered as a 100–200 mg IV bolus over 10 minutes without a test dose. Serious adverse effects are rare with iron sucrose therapy.

Iron replacement therapy

- Intravenous iron is preferred (see Protocol 2.2A).
- Oral iron often fails to improve iron stores because of poor absorption. It is rarely effective in HD patients. Oral iron is frequently associated with gastrointestinal side-effects of diarrhoea or constipation. Phosphate binders may further interfere with the absorption of oral iron.
- IM iron should be avoided because of the risk of cutaneous staining/tattoo and medicolegal concerns due to the risk of neuropraxia.

Epoetin therapy

Erythropoietin refers to the endogenous hormone and epoetin alfa/darbepoetin alfa to the exogenous supplementary erythropoietin preparations. It is imperative to exclude iron deficiency prior to commencing epoetin therapy. Up to 40% of predialysis patients may show a significant increase in Hb concentration following intravenous iron infusion.

Epoetin alfa

- Administered intravenously to HD patients (not licensed for subcutaneous use in Australia due to an association with pure red cell aplasia).
- Usual dose is 50 U/kg per week intravenously.

Darbepoetin alfa

- Licensed for both subcutaneous and IV use.
- Conversion dose from epoetin alfa to darbepoetin alfa is reported to be 200 U:1 μg.
- At higher doses the conversion factor approaches 250–300 U:1 μg.
- Efficacy given by IV or subcutaneous route may be the same.

- Initial dose 0.45 μg/kg once weekly.
- Can be administered fortnightly with only minimal reduction in dosage efficacy.
- Effective in >90% of patients with uraemic anaemia.

Epoetin hyporesponsiveness

Epoetin hyporesponsiveness is defined by:

- a poor response to doses greater than 200 U/kg per week of epoetin alfa
- a poor response to doses greater than 1 μg/kg per week of darbepoetin alfa
- an Hb rise of <15 g/L over a 2-month period

For assessment of patients with a poor response to epoetin see Protocol 2.2B.

The most common cause of epoetin resistance is 'functional iron deficiency' (i.e. ferritin levels may be normal or high but there is a block in release of iron from stores). The TSAT is <20%. Other causes are:

PROTOCOL 2.2B

How to manage poor epoetin responsiveness

- Assess whether the epoetin dose is adequate and administration is correct.
- Assess for haematinic deficiency due to deficiency of iron, vitamin B_{12} or folic acid.
- For iron deficiency give IV iron polymaltose, or iron sucrose if the patient has had a previous adverse reaction. Replace vitamin B_{12} or folic acid if deficient.
- Assess for underlying causes of iron deficiency (absolute or functional) or other haematinic deficiency.
- Assess the adequacy of the patient's dialysis and whether an inflammatory condition is present.
- Exclude severe hyperparathyroidism (osteitis fibrosa), aluminium toxicity or malignancy.

- chronic inflammatory states due to:
  - recurrent infections
  - possibly impure dialysate or bioincompatible dialysis membranes
  - connective tissue diseases
- postoperative state
- inadequate dialysis
- other haematinic deficiencies
- severe secondary hyperparathyroidism
- malignancy
- aluminium toxicity
- pure red cell aplasia secondary to anti-erythropoietin antibodies
- obesity, when epoetin is administered subcutaneously

References

1. Charytan C, Schwenk MH, Al-Saloum MM, Spinowitz BS. Safety of iron sucrose in hemodialysis patients intolerant to other parenteral iron products. Nephron Clin Pract 2004; 96(2): C63–6.
2. CARI (Caring for Australians with Renal Impairment) guidelines. Online. Available: <http://www.kidney.org.au/cari>.
3. Revised European Best Practice Guidelines for the Management of Anaemia in Patients with Chronic Renal Failure. Nephrol Dial Transplant 2004; 19 (suppl 2): 1–47.
4. Macdougall IC, Gray SJ, Elston O, et al. Pharmacokinetics of novel erythropoiesis stimulating protein compared with epoetin alfa in dialysis patients. J Am Soc Nephrol 1999; 10(11): 2392–5.

Chapter 2.3
Bleeding and thrombosis

HIGHLIGHTS

- Both bleeding and thrombosis are common problems in patients on dialysis.
- Bleeding is primarily associated with platelet–platelet and platelet–vessel wall abnormalities. While platelet numbers are generally normal, accumulated evidence has shown acquired defects in platelet glycoprotein (GP) receptors IIb/IIIa for fibrinogen and GPIb for von Willebrand factor (vWF). Anticoagulation therapy during HD and the frequent use of antiplatelet agents exacerbates this bleeding tendency.
- The bleeding time and urea level are crude guides to the potential for bleeding. Responsible uraemic factors have not been fully identified but may include PTH and guanidinosuccinic acid.
- Factors contributing to thrombosis in patients who are on dialysis include hypotension, hyperhomocysteinaemia, endothelial dysfunction, inflammation and malnutrition. An association of increased thrombosis and PD may be influenced by a loss of albumin in the peritoneal dialysate (similar to the thrombotic tendency seen in the nephrotic syndrome).

Conditions that exacerbate bleeding in patients on dialysis

- Inadequate dialysis
- Anaemia: Hb <90 g/L

- Thrombocytopenia due to:
  - immune diseases (e.g. SLE)
  - drugs such as mycophenolate, azathioprine, cyclophosphamide or drug-induced immune thrombocytopenia (see HIT below)
  - bone marrow disorders
- Coagulopathy as a result of:
  - heparin administration during HD
  - increased half-life of low molecular weight heparins resulting in drug accumulation
  - sepsis (disseminated intravascular coagulation)
  - post-plasmapheresis loss of clotting factors

Heparin-induced thrombocytopenia

HIT is a serious adverse effect of heparin seen in patients on dialysis. When heparin binds to platelet factor 4 (H–PF4), the shape of PF4 is altered and it becomes immunogenic. IgG antibodies to H–PF4 can form immune complexes (composed of IgG antibodies and H–PF4) that bind to platelet Fc receptors, causing platelet activation, aggregation, thrombocytopenia, thrombin generation and arterial and venous thromboses. HIT plus the thrombotic syndrome is designated HITTS. It occurs in 1–2% of patients treated with heparin for more than 4 days.

In patients with HIT, 20–50% may go on to develop HITTS after heparin therapy is ceased.

Even with treatment the condition has a mortality of about 20%. Limb loss occurs in about 20% and patients suffer myocardial infarction, stroke and pulmonary embolus.

Unfractionated heparins induce HIT more than low-molecular-weight heparins (LMWHs). However, LMWHs should be used with caution in patients with CKD (see below).

PROTOCOL 2.3A

How to manage serious acute bleeding in patients on dialysis

- Stop HD and resuscitate the patient.
- Collect blood for cross-match and coagulation studies.
- For patients on warfarin therapy, urgent treatment is FFP—for an average adult (80 kg body weight), 4 units FFP is required to produce a 10% increase in coagulation factors. Vitamin K (phytomenadione) 10–20 mg slowly IVI may be effective in 3 hours and oral vitamin K in 4–6 hours.
- For patients with a platelet count $<50 \times 10^9$/L, platelet transfusion is indicated. For intermediate platelet counts of 50–100 $\times 10^9$/L, platelet transfusion is based on the risk of continued significant bleeding and the presence or absence of other treatable risk factors. Transfusion of one single donor apheresis platelet concentrate (equivalent to 4–6 units of pooled platelets) to a non-immunised adult increases the platelet count by approximately 25×10^9/L.
- For patients receiving heparin, this can be reversed using protamine at a dose of 1 mg IV for every 100 units of heparin. The maximum bolus dose is 50 mg over 10 minutes with the remainder given as an IV infusion over 8 hours.
- For patients likely to have platelet dysfunction (particularly if poorly dialysed or previously undialysed) use desmopressin (DDAVP). DDAVP mobilises factor VIII and tissue vWF from endothelial cells. (See Protocol 2.3B.)
- Correct anaemia by transfusion aiming for a Hb level $\geq$90 g/L.
- Raising the haematocrit to 0.3 or above reduces the bleeding time.
- For potential transplant recipients give blood products through a leucocyte filter. Use CMV-matched blood if time permits. *Note:* During non-heparin dialysis, administration of blood though the dialysis circuit is difficult due to clotting.

- For the prevention of further bleeding in uraemic patients, increase the dialysis dose or initiate dialysis.
- For patients with significant bleeds, dialysis options include low-dose or non-heparin HD, regional sodium citrate anticoagulation, the use of prostacyclin or PD (check local protocols).
- Conjugated oestrogens (2.5 mg orally for 5 days) may achieve prolonged control of the bleeding time beginning 1 week after treatment commences and lasting up to 2 weeks (the mechanism is not known). Doses up to 25 mg IVI have been used. Transdermal oestrogens are reported to be safe and effective.

PROTOCOL 2.3B

How to administer desmopressin

- Give IV DDAVP 0.3 μg/kg in 50 mL of normal saline.
- The plasma half-life for factor VIII is 5–8 hours and for vWF is 8–10 hours. The benefits last only a few hours.
- DDAVP may cause vasodilation but hypotension is reduced by infusing over 30 minutes.
- Repeat doses should not be given for 12 to 24 hours after the initial dose.
- The effect diminishes with repeated dosing.

PROTOCOL 2.3C

How to manage post-dialysis fistula bleeding

- Treat with local pressure ± ice packs. *Note:* Excessive pressure may cause fistula thrombosis.
- If bleeding is unusually prolonged or severe, give DDAVP or reverse anticoagulants.

PROTOCOL 2.3D

How to manage bleeding from vascular catheters

- Treat with local pressure ± ice packs.
- Suture bleeding spots if appropriate.
- If bleeding is unusually prolonged or severe, give DDAVP or reverse anticoagulants.
- Watch for compromise of the airways from large neck haematomas and for haemothorax.
- Check FBC serially.
- Admit patient for observation.

Diagnosis

- This is primarily clinical because the available tests are not completely sensitive, specific or rapidly available.
- Tests for the H–PF4 antibody can be classified as functional (demonstrating platelet activation) or immunoassays, which detect antibody but not its functional significance. These tests are time consuming and expensive and immunoassays lend themselves more to batching than to individual testing. However, they may be useful for confirmation of a clinical diagnosis. Only a minority of patients who develop HIT antibodies develop HIT or HITTS.

Treatment

- Stop heparin and commence an anticoagulant that does not cross-react with the H–PF4 antibody. Danaparoid sodium, a synthetic heparinoid with antifactor-X activity, has been used for the treatment of HIT and HITTS. Although uncommon, danaparoid can cross-react with the H-PF4 antibody. (See Protocol 3.5 for HD prescription and monitoring of danaparoid sodium.)
- The direct thrombin inhibitor lepirudin (originally extracted from leeches but now derived by recombinant technology from yeast cells) should not be used because it accumulates in patients with renal impairment.

- Argatroban does not accumulate in patients with renal failure but it is unavailable in Australia.

Low-molecular-weight heparins (anti-Xa)

- Dalteparin and enoxaparin sodium are safe and effective for anticoagulation during chronic HD. Protocol 3.5 provides dose guidelines for LMWHs when used for HD anticoagulation.
- However, in patients with CKD or on dialysis, LMWHs should be used cautiously, if at all, for the treatment of DVT, unstable angina or non-Q-wave myocardial infarction.
- The clearance of LMWHs is reduced in renal disease and in elderly patients. Therefore, unpredictable anticoagulant effects and life-threatening bleeding can occur if LMWHs are administered at fixed-weight doses and without monitoring.
- If used, the dose of the LMWH should be reduced, the dose interval increased and anti-Xa activity should be monitored.
- Conventional unfractionated heparin is first-line therapy.

References

1. Salvati F, Liani M. Role of platelet surface receptor abnormalities in the bleeding and thrombotic diathesis of uremic patients on hemodialysis and peritoneal dialysis. Int J Artif Organs 2001; 24(3): 131–5.
2. Noris M, Remuzzi G. Uremic bleeding: closing the circle after 30 years of controversies? Blood 1999; 94(8): 2569–74.
3. Hirsh J, Heddle N, Kelton JG. Treatment of heparin-induced thrombocytopenia: a critical review. Arch Intern Med 2004; 164(4): 361–9.
4. Janssen MJ, van der Meulen J. The bleeding risk in chronic haemodialysis: preventive strategies in high-risk patients. Neth J Med 1996; 48(5): 198–207.

Chapter 2.4
Parathyroid hormone, vitamin D, calcium, phosphate and bones

HIGHLIGHTS

- Most patients have abnormal bone histology by the time they commence dialysis as a result of abnormal bone turnover and/or mineralisation.
- Assays for intact-PTH (iPTH), biointact-PTH (1–84) and bone-specific alkaline phosphatase (BSAP) are guides to the rate of bone turnover but may not reflect bone histomorphometry.
- Hyperphosphataemia may reflect 'inadequate' dialysis. An increased dialysis period (e.g. short daily or overnight dialysis) will improve phosphate levels (sometimes with hypophosphataemia requiring supplementation).
- When using calcium-based phosphate binders, avoid hypercalcaemia and PTH suppression that can lead to low bone turnover.
- Achieving biochemical targets for calcium, phosphate and PTH levels may reduce the risk of bone disease and serious cardiovascular complications.
- Parathyroidectomy should be considered early in patients resistant to medical management.

Bone histology in dialysis patients

The patterns of bone histology are:

- high-turnover bone disease: present in about 50% of patients (called osteitis fibrosa if severe), caused by excessive PTH release

- osteomalacia: present as the sole change in about 5–10% of patients, indicating inadequate mineralisation of osteoid bone matrix
- low-turnover bone disease: present in 20–25% of patients (called adynamic bone disease if severe) and generally associated with suppression of iPTH and BSAP
- mixed pathology: present in about 10–15% of patients, usually comprising high-turnover bone disease plus abnormal mineralisation

These abnormal patterns of turnover and mineralisation cause a deterioration of bone architecture, often accompanied by low BMD, so that patients on dialysis are prone to fracture.

High-turnover, hyperparathyroid bone disease

Pathogenesis

- Levels of active vitamin D (1,25-dihydroxy cholecalciferol/ calcitriol) fall. Calcitriol acts via vitamin D receptors in the parathyroid glands to suppress PTH production. As calcitriol levels fall, PTH levels increase.
- Calcium levels may fall due to a reduction in calcitriol-mediated uptake of calcium from the gastrointestinal tract. Ionised calcium levels normally inhibit PTH release by signalling through extracellular calcium-sensing receptors on parathyroid cells. If this signal is reduced, PTH release is increased.
- Increased serum phosphate levels reduce ionised calcium levels and act directly on the parathyroid glands to increase PTH.
- Parathyroid cell proliferation occurs over time. With hyperplasia and nodular (monoclonal) hyperplasia, the basal level of PTH production increases. The parathyroid glands may also become less responsive to calcium and vitamin D feedback due to reduced calcium receptor and vitamin D receptor expression.

Clinical features

- PTH stimulates osteoblast and osteoclast activity, bone turnover and the release of calcium and phosphate from bone. Abnormal serum calcium and phosphate levels contribute to pruritus and subcutaneous and vascular calcification.
- Sustained PTH elevation leads to abnormal bone architecture, bone pain and increased fracture risk.

Radiographic features

- Subperiosteal erosion of digital phalanges, distal clavicles, ischium, pubis and SI joints. Brown tumours—areas of local bone erosion and fibrosis.
- Patchy osteosclerosis may produce 'rugger jersey' appearance on lateral X-rays of spine.

Osteomalacia

OM is a disorder of decreased bone mineralisation that is best diagnosed by bone biopsy with double tetracycline labelling.

Pathogenesis

- Hypovitaminosis D.
- Acidosis.
- May accompany high- or low-turnover bone disease (high-turnover bone disease + mineralisation defect = mixed osteodystrophy).

Clinical features

- May be asymptomatic.
- Bone or muscle pain.
- Low trauma fracture.

Radiographic features

- Pseudofractures or Looser's zones: radiolucent bands oriented perpendicular to the long axis of cortical bones.

Low-turnover bone disease

LTBD (or adynamic bone disease when bone turnover is absent) is usually associated with levels of iPTH that are less than twice the assay upper range but may be present on bone biopsy when iPTH levels are at or above the target range.

Pathogenesis

- Bone morphogenic protein-7 (BMP-7) is a bone anabolic agent that is important for osteoblast development. Recently, it has been suggested that patients with CKD have reduced BMP-7 production. This could reduce bone turnover, leading to a compensatory increase in PTH secretion. However, excessive suppression of PTH by calcium or vitamin D therapy can return the patient to a state of low bone turnover.
- Diabetes and old age are associated with low bone turnover.
- Aluminium exposure can cause low bone turnover by reducing PTH secretion, osteoblast activity and crystal growth.
- PTH resistance may occur due to reduced PTH receptor sensitivity or expression and to the opposing effect of long C-terminal PTH fragments.

Clinical features

- Fracture (inadequate repair of bone damage).
- Calcium levels may be elevated (reduced bone uptake of excess calcium).
- Increased cardiovascular risk.

Osteoporosis

- Patients on dialysis often have a reduced BMD.
- Osteoporosis (defined by a BMD of more than 2.5 SD below the mean BMD of normal young adults) is more common in elderly patients, women and patients with type 1 diabetes.
- A low BMD contributes to the increased risk of low trauma fracture.

- Abnormal bone architecture due to low bone turnover, hyperparathyroidism and abnormal mineralisation also contribute to fracture risk.

Diagnosis of bone disease

Intact-PTH and BSAP assays

Renal bone disease is an abnormality of bone turnover and mineralisation. Intact-PTH assays and alkaline phosphatase or BSAP levels are often used to estimate bone turnover.

Based on correlation with histomorphometry:

- iPTH levels less than twice the assay upper range generally indicate low bone turnover
- iPTH levels of 2–3 times the assay upper range generally indicate normal bone turnover but are not always predictive
- iPTH levels more than 7 times the assay upper range predict high bone turnover or normal bone

Intact-PTH assays measure both PTH (1–84) and long carboxyl-terminal fragments, including PTH (7–84). Less data are available for newer 'biointact' PTH (1–84) assays but they appear to predict bone histomorphometry with similar accuracy to assays for iPTH.

Bone biopsy

The 'gold standard' for diagnosis of bone disease is bone biopsy. Because more than one pathology is often present, bone biopsy (Protocol 2.4A) assists in answering the following questions:

- Does this patient have bone disease due to aluminium exposure?
- What is the cause of this patient's hypercalcaemia?
- Is desferrioxamine chelation therapy required?
- Is osteomalacia present?
- What factors are contributing to the worsening of this patient's BMD?

PROTOCOL 2.4A

How to prepare patients for a bone biopsy

- Prescribe tetracycline hydrochloride 250 mg 3 times daily on days 1 and 2.
- Do not take tetracycline for the next 10 days.
- Prescribe tetracycline hydrochloride 250 mg 3 times daily on days 13 and 14.
- The bone biopsy is performed 3–5 days after the second tetracycline label, usually from the iliac crest.
- Requirements: size 8 Jamshidi disposable bone biopsy needle, midazolam, 1% lignocaine for local infiltration, scalpel blade and procedure pack.
- Transport specimen in chilled buffered formalin.
- Request undecalcified bone histology and histomorphometry. Liaise with the laboratory beforehand as this type of bone histology is not widely available. When sending the specimen, include a copy of the protocol you have used.
- The patient can leave, accompanied, 1 hour after recovery from premedication if observations are stable.

Bone scans

Bone scans are non-specific. The typical scan of hyperparathyroid bone disease is described as a 'super scan' with a symmetrical increase of tracer activity.

Reduced BMD and osteoporosis are diagnosed using dual energy X-ray absorptiometry (DXA) or quantitative CT (QCT).

Aluminium toxicity

Routine screening of plasma aluminium levels may indicate patients at risk of toxicity. Protocol 2.4B indicates how to interpret plasma aluminium levels and Protocol 2.4C how to perform a desferrioxamine (DFO) challenge prior to consideration of treatment. Protocol 2.4D deals with how to commence treatment with DFO.

PROTOCOL 2.4B

How to interpret plasma aluminium levels

- Most information indicates that plasma Al levels only reflect recent Al intake.
- Monitoring at once or twice yearly intervals is useful to detect exposure.
- Al levels should be <2 μmol/L (50 μg/L).[a]
- IV iron supplements may reduce basal plasma Al levels. When transferrin saturation levels are >60%, low serum Al levels may be associated with Al overload and Al-related bone disease.
- When the plasma Al level is <2 μmol/L, a DFO challenge should be considered if low-turnover bone disease, including osteomalacia, neurological disease (other than peripheral neuropathy) or anaemia (especially microcytic, hypochromic), is present or considered likely.
- When the plasma Al level is >2–3.5 μmol/L, identify the Al source and consider a DFO challenge, particularly in the presence of bone disease, neurological disease or anaemia.
- When the plasma Al level is >3.5 μmol/L, the risk of Al-related bone disease is increased and a DFO challenge should be performed.
- To avoid serious side-effects, withhold the DFO challenge until the plasma Al level is <4.5 μmol/L. Withdraw all sources of Al exposure (drugs and dialysate) and for levels over 7.5 μmol/L consider daily high-flux dialysis.

(*a*) Divide plasma Al in μmol/L by 0.0371 to convert to μg/L.

PROTOCOL 2.4C

How to perform a low-dose desferrioxamine challenge

- Measure plasma Al level prior to the commencement of dialysis (pre).
- Give 5 mg/kg IV DFO during the last hour of the dialysis.[a]
- Measure plasma Al level prior to the next HD 2 days later (post).
- The challenge is positive when the increment in Al level is >2 μmol/L (50 μg/L) and highly positive when the rise is >11 μmol/L (300 μg/L).
- The reliability of a positive DFO challenge is improved when the iPTH level is less than 2–3 times the normal upper range of the assay.
- Adynamic Al-related bone disease may be present in 85% of patients with a positive DFO challenge and iPTH less than 2–3 times the assay's upper normal range.
- Perform a bone biopsy if the DFO challenge is positive.
- Bone biopsy is considered positive for Al when Al covers 15–25% of the bone surface. This is often associated with reduced bone formation and osteomalacia.

(*a*) Higher (20–40 mg/kg) DFO doses are no longer used due to the risk of visual loss and mucormycosis. DFO chelates iron to form ferrioxamine, which can enhance the growth of several species of *Mucor*.

PROTOCOL 2.4D

How to commence treatment with desferrioxamine after a positive bone biopsy

- Check the local protocol. One method is to give 5 mg DFO IV once weekly during the last hour of dialysis for 4 months.
- Before giving each dose ask about side-effects, which include: visual impairment, auditory neurotoxicity (dose dependent and partly reversible), interaction with antipsychotic agents, yersinosis and other unusual infections, allergic reactions, hypotension and iron deficiency.
- LFTs, ferritin and transferrin saturation should be checked at usual 2-monthly intervals.
- Plasma Al levels and visual acuity should be tested every 3 months.
- Repeat low-dose DFO challenge at 4 months. If the plasma Al level after DFO is:
  - >11 μmol/L (300 μg/L), continue DFO for a further 4 months
  - 2–11 μmol/L (50–300 μg/L), continue DFO for a further 2 months
  - <2 μmol/L (50 μg/L), cease treatment for 1 month then repeat the low-dose DFO challenge.
- Repeat the bone biopsy at the conclusion of DFO treatment.

Biochemical targets

Biochemical targets have been established for calcium, phosphate and the calcium by phosphate product and targets have been set for iPTH (Table 2.4). If these targets are achieved, renal bone disease and cardiovascular complications of abnormal calcium and phosphate regulation are less likely to progress.

Table 2.4 Preferred targets for calcium, phosphate, intact-PTH and the calcium by phosphate product

	CARI[a]		K/DOQI[b]	
	CKD stage 3–4	**CKD stage 5**	**CKD stage 3–4**	**CKD stage 5**
Serum calcium (mmol/L)	Normal range	Normal range	Normal range	2.1–2.37
Serum phosphate (mmol/L) (predialysis mid-week for HD)	Normal range	<1.8	0.87–1.48	1.13–1.78
Intact-PTH (pmol/L)	2–4 fold	2–3 fold	Normal–2 fold	2–4.5 fold
Calcium by phosphate product $(mmol/L)^2$		<4.2		<4.4

(*a*) CARI (Caring for Australians with Renal Impairment). Online. Available: <http://www.kidney.org.au>.
(*b*) K/DOQI. Kidney Disease Outcomes Quality Initiative. Am J Kidney Dis 2003; 42(4) (suppl 3): S53, S63, S77.

Drugs used in the management of phosphate, calcium and PTH disturbance

Phosphate binders

- These drugs form insoluble complexes with dietary phosphate and should be taken just before or with food.
- Calcium carbonate may cause hypercalcaemia.
- Magnesium salts are used (measure Mg; may require reduction of Mg in dialysate).
- Aluminium hydroxide is used (prolonged use may cause Al toxicity).
- Sevelamer (a cationic exchange resin, which also reduces LDL-C) and lanthanum carbonate (a rare earth metal) are non-calcium or aluminium-containing phosphate binders. These drugs are expensive and their availability is currently restricted.

Calcitriol/active vitamin D analogues

- These suppress PTH production. They may cause hypercalcaemia, increase phosphate absorption and increase the calcium by phosphate product.
- They are best used before the development of severe hyperparathyroidism.
- They may be given daily or as pulse therapy (weekly or 2–3 times weekly).

Calcimimetics

- These compounds enhance the response of the extracellular parathyroid calcium-sensing receptor to ionised calcium.
- This enhanced response suppresses PTH release and reduces serum phosphate and calcium levels.
- The use of cinacalcet will be limited by cost and its availability is currently restricted.

Bisphosphonates

These drugs reduce osteoclastic bone resorption and hence serum calcium and phosphate levels. They are available as oral and IV preparations.

Uses

- Generally the use of IV bisphosphonates (e.g. pamidronate and zolendronic acid) is restricted to hypercalcaemia, severe hyperphosphataemia and calcific vasculopathy (calciphylaxis).
- They may be useful while awaiting parathyroidectomy.
- They may improve BMD by reducing osteoclastic bone resorption, allowing time for resorption pits to be filled.

Morbidity and mortality from complications of secondary hyperparathyroidism and vascular and soft tissue calcification may be reduced by careful control of the serum phosphate levels. Protocols 2.4E, 2.4F and 2.4G describe the management approaches to hyperphosphataemia in different clinical settings.

PROTOCOL 2.4E

How to manage hyperphosphataemia when intact-PTH levels are below the target range and low bone turnover is likely

- Reduce or cease calcitriol.
- Increase dialysis dose/adjust diet (see Table 2.10B).
- Use a non-calcium phosphate binder. Sevelamer or lanthanum are best suited but expensive/not readily available. Use a magnesium-based phosphate binder that does not contain aluminium.
- Consider using low calcium dialysis concentration plus a calcium-based binder.
- If an inappropriately low iPTH is accompanied by hypercalcaemia, exclude aluminium overload (plasma Al and DFO challenge/bone biopsy).

PROTOCOL 2.4F

How to manage hyperphosphataemia when intact-PTH levels are in the target range

- Increase dialysis dose/adjust diet (see Table 2.10B).
- Reduce or cease calcitriol.
- Use calcium carbonate as the first-line treatment.
- Aluminium- and magnesium-based binders can be used to supplement calcium.
- Sevelamer or lanthanum are well suited but expensive/not readily available. Consider their use in compliant younger patients who require high doses of calcium-based binders.

PROTOCOL 2.4G

How to manage hyperphosphataemia when intact-PTH levels are above the target range

- Increase dialysis dose/adjust diet (see Table 2.10B).
- Reduce or cease calcitriol.
- If the patient is normocalcaemic, calcium carbonate is the first-line treatment, with the addition of aluminium- and magnesium-based binders if necessary.
- If the patient is hypercalcaemic and the iPTH is in the range of 4–7 times the normal upper range, cinacalcet is appropriate but expensive/not readily available. Aluminium- and magnesium-based binders can be used. Sevelamer and lanthanum are suitable but expensive/not readily available.
- IV bisphosphonate treatment (e.g. pamidronate 1 mg/kg as a slow IV infusion) can be used for short-term control of severe hyperphosphataemia, particularly if it is accompanied by hypercalcaemia.
- Parathyroidectomy should be considered early in patients with severe hyperparathyroidism (iPTH >7–10 times normal upper range with hypercalcaemia) who are unresponsive to the above medical management.

PROTOCOL 2.4H

How to manage patients pre- and post-parathyroidectomy to reduce complications of hypocalcaemia

- If the patient is not hypercalcaemic, calcitriol can be commenced preoperatively at a dose of 2 μg (8 capsules) daily for 3 days.
- Alternatively, give calcitriol 2 μg on the preoperative night.
- Ensure preoperative serum magnesium level is ≥0.80 mmol/L. Supplement with Mylanta or magnesium aspartate if below this level.
- Patients with severe hyperparathyroidism (e.g. ALP >350 U/L, iPTH 10–15 times normal) will often require postoperative IV calcium. Arrange central venous access intraoperatively in these patients.
- Measure the serum total calcium or preferably ionised calcium level every 4–6 hours postoperatively for 48–72 hours.
- Commence calcium gluconate (1–2 mg/kg/h) if the serum calcium level is <1.8 mmol/L or ionised calcium level is <0.9 mmol/L (10 mL of 10% calcium gluconate contains 90 mg of elemental calcium). Adjust the dose to maintain the total or ionised calcium level in the normal range. Reduce the dose when the serum calcium level is stable and in the normal range.
- Commence patients on oral calcium (1–2 g 3 times daily) and calcitriol (up to 2 μg daily) as soon as possible.

Parathyroidectomy

This is indicated when PTH is elevated and associated with:

- severe hypercalcaemia (serum total calcium ≥3 mmol/L) after calcium and calcitriol withdrawal
- progressive hyperparathyroid bone disease
- worsening osteoporosis
- calcific vasculopathy

Parathyroidectomy may also improve hyperphosphataemia, pruritus, myopathy and anaemia.

After parathyroidectomy the serum calcium and phosphate levels may fall abruptly ('hungry bone syndrome'). Protocol 2.4H details pre- and post-parathyroidectomy management to reduce hypocalcaemia.

Treatment of osteoporosis

- Patients on dialysis are often elderly and unwell, poorly nourished and may have visual impairment and peripheral neuropathy causing impaired postural stability. These factors increase the risk of osteoporotic fracture.
- Abnormal bone turnover and mineralisation due to renal osteodystrophy further worsen bone quality.
- Although the treatment of osteoporosis diagnosed by DXA (or QCT) cannot be separated from the treatment of osteodystrophy, standard osteoporosis treatment options, including oral bisphosphonates (possibly at reduced dosage), IV bisphosphonates, selective (o)estrogen receptor modulators (SERMS) and possibly hormone replacement therapy (HRT) are applicable to some patients.

References

1. Elder G. Pathophysiology and recent advances in the management of renal osteodystrophy. J Bone Miner Res 2002; 17(12): 2094–105.
2. D'Haese PC, Couttenye MM, De Broe ME. Diagnosis and treatment of aluminium bone disease. Nephrol Dial Transplant 1996; 11 (suppl 3): 74–9.
3. van Landeghem GF, D'Haese PC, Lamberts LV, et al. Low serum aluminium values in dialysis patients with increased bone aluminium levels. Clin Nephrol 1998; 50(2): 69–76.
4. CARI. Caring for Australians with Renal Impairment. Online. Available: <http://www.kidney.org.au>.
5. K/DOQI. Kidney Disease Outcomes Quality Initiative. Am J Kidney Dis 2003; 42(4) (suppl 3): S53, S63, S77, S108, S116, S127.
6. Hruska KA, Saab G, Chaudhary LR, et al. Kidney–bone, bone–kidney, and cell–cell communications in renal osteodystrophy. Semin Nephrol 2004; 24(1): 25–38.

Chapter 2.5
Endocrine disturbances of chronic kidney disease

HIGHLIGHTS

- In addition to hormonal disturbance of erythropoiesis (Chapter 2.2) and bone and mineral metabolism (Chapter 2.4), abnormalities of gonadotrophins, sex steroids, prolactin and thyroid hormones are common in men and women who are on dialysis.
- Causes include:
  - altered hypothalamic–pituitary axis function and feedback responsiveness
  - reduced end-organ responsiveness (e.g. reduced Leydig cell responsiveness to LH)
  - reduced metabolic clearance rates (e.g. insulin and possibly prolactin)

Common hormonal changes in males with CKD

These are outlined in Table 2.5A.

Common hormonal changes in females with CKD

These are outlined in Table 2.5B.

Hormonal changes common to both genders

These are outlined in Table 2.5C.

Table 2.5A Common hormonal changes in males with chronic kidney disease

Hormone	Effect of CKD on hormone levels	Clinical features	Treatment[a]
Testosterone LH FSH	Normal or reduced Normal or increased Normal or increased	• Hypogonadism • Reduced libido and sexual function • Increased body fat and reduced lean body mass • Reduced bone mineral density	• Testosterone[b] • Erythropoietin • Zinc may be useful in zinc-deficient individuals
Prolactin	Generally not greater than 5 times the assay normal upper range	• Hypogonadism • Reduced libido and sexual function	• Treat hyperparathyroidism or hypothyroidism if present • Erythropoietin • Consider trial of dopamine agonists in symptomatic patients with hyperprolactinaemia
Oestradiol	Typically normal		• Not applicable

(*a*) Individualised based on investigations and clinical features.
(*b*) Refer to drug literature before prescribing IM testosterone enanthate or esters, subdermal implants or transdermal patches. Prostate cancer is an absolute contraindication to use. Androgens may influence clotting factors and the lipid profile.

Table 2.5B Common hormonal changes in females with chronic kidney disease

Hormone	Effect of CKD on hormone levels	Clinical features	Treatment[a]
Premenopausal females			
Oestradiol LH FSH	Normal or reduced Normal or increased Normal or increased	• Oligomenorrhea or amenorrhoea • Reduced libido • Reduced bone mineral density	• Improve nutrition and body weight • Erythropoietin • Gynaecological assessment advised. Progestogen if evidence of unopposed oestrogen effect on endometrium, trial of oestrogen or HRT.
Postmenopausal			
Oestradiol, FSH and LH	In normal postmenopausal range	• As above	• HRT for symptom relief ± treatment of low BMD
Pre-/postmenopausal			
Prolactin	Generally not greater than 5 times the assay normal upper range	• As above ± galactorrhoea	• Treat hyperparathyroidism or hypothyroidism if present • Erythropoietin • Consider trial of dopamine agonists in symptomatic patients with hyperprolactinaemia

(*a*) Individualised based on investigations and clinical features.

Table 2.5C Hormonal changes common to both genders

Hormone	Effect of CKD on hormone levels	Clinical features	Treatment[a]
Total T_3 and free T_3 TSH, free T_4 and reverse T_3	Generally reduced Generally normal. TSH, T_3 and T_4 may be below normal during severe illness	• Most patients are euthyroid	• Not applicable
SHBG	No change in uraemia Increase with ageing	• Not applicable	• Not applicable
Insulin	Insulin resistance and hepatic insulin metabolism improve with dialysis	• Uraemia may induce overt diabetes but effects are variable	• Therapy must be individualised

(*a*) Individualised based on investigations and clinical features.

Some drugs that commonly influence hormone responses

- Glucocorticoids reduce testosterone biosynthesis via gonadal steroid receptors and at the hypothalamus–pituitary axis.
- Oestrogen sensitises lactotroph cells to stimuli that induce prolactin secretion.
- Cimetidine and spironolactone block androgen receptors.
- Spironolactone and ketoconazole reduce 17α-hydroxylase activity, leading to reduced testosterone biosynthesis.
- Cyclosporin and tacrolimus may have direct effects on Leydig cells.
- Amiodarone, an iodine-rich drug, may precipitate hypothyroidism or hyperthyroidism.
- Conventional antipsychotic drugs and some atypical antipsychotics (risperidone, amisulpride) block dopamine receptors and release prolactin from dopaminergic inhibition.

References

1. Palmer BF. Sexual dysfunction in uremia. J Am Soc Nephrol 1999; 10(6): 1381–8.
2. Schmidt A, Luger A, Horl WH. Sexual hormone abnormalities in male patients with renal failure. Nephrol Dial Transplant 2002; 17(3): 368–71.

Chapter 2.6
Infection control

HIGHLIGHTS

- The underlying principle of infection control is the recognition that patients can acquire or spread infections while receiving health care.
- These infections can be minimised by adopting quality control measures.
- This section outlines common infection control procedures.

Infection screening before and after the commencement of dialysis

- Patients should undergo screening for common and communicable infections prior to commencing dialysis, whether or not they are classified as being in a group at 'high risk'.
- Until results are available, HD machines used by these patients should be disinfected after each run with a rapid and long-acting bactericidal agent (e.g. Dialox).
- Table 2.6A indicates initial and follow-up screening procedures.

Precautions taken during the dialysis of patients with infection

The standard infection control precautions to be taken during the dialysis of patients with infection are outlined in Table 2.6B.

Table 2.6A Infection surveillance regimen for patients commencing or continuing dialysis

Initial test	Result	Vaccination	Retesting frequency (months)
HbsAg and anti-HBs	Negative	Yes	6
anti-HBs	Positive (>10 mIU/mL)	No	6
HbsAg	Positive	No (consultation regarding treatment)	
anti-HCV	Negative	Not applicable	6
anti-HCV	Positive	Not applicable (consultation regarding treatment)	12
HIV	Negative	Not applicable	12
	Positive	Not applicable (consultation regarding treatment)	
MRSA	Negative or positive	Not applicable	6
VRE (rectal swab)	Negative or positive	Not applicable	6

anti-HBs: hepatitis B surface antibody; anti-HCV: hepatitis C antibody; HbsAg: hepatitis B surface antigen; HIV: human immunodeficiency virus; MRSA: methicillin-resistant *Staphylococcus aureus*; VRE: vancomycin-resistant *Enterococcus*.

Table 2.6B Precautions for patients with infection

Infection	Standard precaution (patient and equipment)	Standard precaution (staff)
HBsAg positive[(a)]	Isolation of both patient and machine. Standard clean of equipment	Goggles, gloves and long-sleeved gown
anti-HCV positive	Non-isolation of patients or machine. Dialysis station cleaned after treatment	Goggles, gloves and long-sleeved gown
HIV positive	Non-isolation. Dialysis station cleaned after treatment	Goggles, gloves and long-sleeved gown
VRE positive	Patient and machine isolation. Dialysis station cleaned after treatment	Goggles, gloves and long-sleeved gown when in direct patient contact
TB in potentially infectious patient	Isolated in negative pressure room	Goggles, gloves, particulate mask when in direct patient contact and long-sleeved gown
MRSA positive	Non-isolated unless respiratory involvement	Goggles, gloves and long-sleeved gown
Legionnaires' disease	Non-isolation. Standard precautions.	Goggles, gloves and long-sleeved gown
Meningococcal meningitis	Isolated in negative pressure room	<24 hours: respiratory precautions After antibiotic therapy and clearance: standard precautions

(*a*) HBV DNA is sometimes found in HBsAg negative patients.

Recommended vaccinations for patients on dialysis

For patients with no prior vaccination or protective antibody titre the following vaccinations are recommended:

- hepatitis B (see Protocols 2.6A and 2.6B)
- influenza (yearly)
- pneumococcal polysaccharide vaccine (PPV) (recommended every 5 years)
- tetanus (recommended every 10 years)
- diphtheria (recommended every 10 years)
- varicella vaccine if antibody negative

Note: Chickenpox and measles, mumps and rubella (MMR) vaccines should not be used in immunosuppressed patients or after transplantation.

PROTOCOL 2.6A

How to vaccinate for hepatitis B in patients with chronic kidney disease

Using recombinant HBV vaccine (Engerix-B or H-B-Vax II):

- vaccinate patients who are negative for hepatitis B surface antigen (HBsAg) and hepatitis B surface antibody (anti-HBs)
- if severely uraemic, vaccination can be postponed until 1 month after commencement of dialysis
- the standard regimen is 40 μg or 20 μg at 2 sites, into the thigh or upper arm (this is twice the standard dose for patients without CKD)
- repeat the dose at 1 and 6 months
- check for anti-HBs 1 month after completion of course

PROTOCOL 2.6B

How to manage non-responders/weak responders to hepatitis B vaccine

Anti-HBs levels <10 mIU/mL indicate non-responders or weak responders. Using recombinant HBV vaccine (Engerix-B or H-B-Vax II):

- give an additional 40 μg dose and check response after 1 month

or

- give a further 3 doses at 1 month intervals and check response 1 month later

or

- give 5 μg by intradermal injection every week for a total of 10 doses—this regimen results in a higher conversion rate at 6 months; the duration of immunity does not appear to differ between the IM and intradermal routes

References

1. Chau KF, Cheng YL, Tsang DN, et al. Efficacy and side effects of intradermal hepatitis B vaccination in CAPD patients: a comparison with the intramuscular vaccination. Am J Kidney Dis 2004; 43(5): 910–17.
2. National Kidney Foundation. Online. Available: <http://www.kidney.org>.
3. NHMRC. Australian immunisation handbook. 8th edn. Vaccines: 3.9 Hepatitis B. Canberra: NHMRC, 2003.

Chapter 2.7
Pregnancy and dialysis

HIGHLIGHTS

- Women undergoing dialysis have markedly reduced fertility because of uraemia, anaemia and abnormal regulation of the hypothalamic–pituitary–ovarian axis.
- Although pregnancy occurs infrequently in women undergoing maintenance dialysis, it appears to be increasingly common. This may reflect improved dialysis and the beneficial effects of erythropoietin therapy.
- False positive and negative urinary β-HCG results may occur in patients on dialysis and, because amenorrhoea may already be present, pregnancy may be first diagnosed on ultrasound.
- The risks of pregnancy for both mother and infant should be discussed in detail. For a successful outcome, scrupulous attention must be paid to fluid balance, hours of dialysis, nutrition and blood pressure management.

Dialysis management of the pregnant woman

- For some pregnant women with advanced CKD, dialysis should be initiated to reduce adverse maternal and foetal effects of the uraemic environment.
- Dialysis permits better control of fluid overload, BP and an increased caloric and protein intake.

- Women who conceive before commencing dialysis deliver a viable infant in about 75% of cases.
- HD and PD are safe during pregnancy and women who conceive while established on dialysis deliver a viable infant in about 40% of cases. Patients usually remain on their established dialysis mode. Placement of a PD catheter is possible at any stage of pregnancy.
- Improved maternal and foetal outcomes occur with more than 20 hours per week of HD and urea values <20 mmol/L. Based on limited information, six treatments per week are recommended for patients on HD. This regimen decreases the fluid removal requirement at each treatment and the risk of hypotension, placental hypoperfusion, polyhydramnios and premature labour.
- At least 6–8 exchanges daily are recommended for patients on PD. In late pregnancy it becomes difficult for women to tolerate usual exchange volumes. Daytime CAPD and continuous cycling PD at night can be used to increase the number of exchanges made necessary by a reduction in exchange volumes.

Associated management

- Close monitoring of serum potassium and calcium levels and acid–base balance is required. Adjustment of the dialysate prescription may be necessary.
- Folate requirements are increased and high-dose supplementation (5 mg daily) is recommended.
- Increased erythropoietin doses are often required.
- Iron requirements are increased. IV iron should be given in intermittent doses (e.g. 100 mg Ferrum H weekly) to avoid foetal iron toxicity.

Management of hypertension in a pregnant patient on dialysis

- The major maternal risk is severe HT, generally occurring before the third trimester. There is an increased risk of pre-eclampsia.

- As with non-dialysis patients, only a limited number of antihypertensive drugs are regarded as 'safe' during pregnancy. These include methyldopa, oxprenolol, labetalol, clonidine and hydralazine.

Seizure prophylaxis prior to delivery

Magnesium is superior to other anticonvulsants for seizure prophylaxis in women with pre-eclampsia but it must be used with extreme caution in dialysis patients. Magnesium potentiates the hypotensive effects of calcium channel blockers. While a standard loading dose is considered safe, extreme caution should be used if a magnesium infusion is considered necessary.

Perinatal complications

- Early involvement of a perinatologist and neonatologist is critical to the pregnancy outcome. Most infants born to women on dialysis are premature (mean gestation age, 33 weeks) and are frequently small for gestational age.
- At birth, infants with normal kidney function will nevertheless have urea and creatinine levels similar to those of the mother. The resulting solute diuresis requires careful monitoring of the infant's electrolyte and volume status.

Contraception guidelines

Contraception should be discussed with all potentially fertile patients on dialysis. Women who become pregnant once on dialysis may conceive again. In general, oral contraceptive agents and barrier methods are appropriate, although oestrogens may worsen HT in some patients. IUDs are generally avoided due to the risks of infection and bleeding.

References

1. Okundaye I, Abrinko P, Hou S. Registry of pregnancy in dialysis patients. Am J Kidney Dis 1998; 31: 766–73.

2. Nakabayashi M, Adachi T, Itoh S, et al. Perinatal and infant outcome in pregnant patients undergoing chronic haemodialysis. Nephron 1999; 82: 27–31.
3. Chao A, Huang J, Lien R, et al. Pregnancy in women who undergo long-term haemodialysis. Am J Obstet Gynecol 2002; 187: 152–6.
4. Hou S. Modification of dialysis regimens for pregnancy. Int J Artif Organs 2002; 25: 823–6.

Chapter 2.8
Preparation for transplantation

HIGHLIGHTS

- Numerous reasons exist to support the widespread use of renal transplantation and combined renal–pancreas transplantation. These include improved quality of life, life expectancy, employment prospects, fertility and reduced comorbidity and family stress.
- Transplantation is also a less costly form of renal replacement therapy than dialysis.

Types of organ transplantation in renal failure

- Cadaveric kidney donation.
- Living related kidney donation.
- Living unrelated kidney donation.
- Cadaveric kidney–pancreas donation (recipients with type 1 diabetes).

Cadaveric organ donation

- Cadaveric donation can follow brain death, where the donor is classed as a 'beating heart donor', or irreversible cessation of the circulation, where the donor is classed as a 'non-beating heart donor'.
- 'Warm ischaemia', which commences from the moment blood circulation ceases, can cause acute tubular necrosis and primary 'delayed graft function'. Providing the injury is not too severe, the kidney remains suitable for transplantation.

Cadaveric organ allocation

- In Australia, the Australian Red Cross oversees organ allocation.
- Demand for cadaveric kidneys outstrips supply by 5:1. Organ allocation criteria have been created to optimise transplant outcomes while ensuring impartial allocation.
- After excluding those with ABO blood group incompatibility, all potential organ recipients receive the same base score. Points are deducted for each human leucocyte antigen (HLA) mismatch. The potential recipient on the transplant waiting list with the best HLA match is offered the kidney, provided he or she has a 'negative cross-match'.
- A negative T and B cell cross-match between donor and recipient is required prior to transplantation. In the case of live donors, this is performed on a number of occasions and is repeated 1–2 weeks prior to the proposed transplant.

Live donors

- Live donor renal transplantation has been an accepted form of transplantation since 1954. Due to the disparity between cadaveric donors and recipients, there has been a shift to encourage live donor renal transplantation.
- First-degree relatives are the most likely source of a living HLA-matched donor.
- More recently there has been a shift from living related (parent/child or sibling/sibling) transplants to living unrelated organ donation (spouse or friend).

Risks of 'living related' and 'living unrelated' kidney donation

- Risk of death is 0–0.03%.
- Risk of acute complications is 0.25–2.5% and the most frequent complications (>1%) are similar to those

encountered with any type of surgical intervention, including: pneumonia/atelectasis, pneumothorax, wound/urinary tract sepsis, haemorrhage and thromboembolic disease.
- With careful donor evaluation the long-term risks of living with one kidney are negligible.

Pretransplantation education

- All patients should attend pretransplantatation education. The recipient should be introduced to other transplant recipients if possible.
- Education sessions should outline the relative merits and complications of transplantation, particularly the risks of sepsis and malignancy, and evaluate the patient's understanding and commitment to long-term compliance with immunosuppression medication.

Protocols 2.8A–2.8F address the assessment of donors and recipients.

Preoperative and postoperative dialysis

- Prior to transplantation a short run of HD is usually required to correct fluid overload or hyperkalaemia.
- Heparin should not be used.
- Avoid preoperative dehydration. Aim for approximately 1 kg above the patient's usual dry weight.
- Primary non-function (failure to provide sufficient renal function to avoid dialysis within 48 hours of transplantation) occurs in approximately 15% of donor kidneys.

PROTOCOL 2.8A

How to assess patients on dialysis for transplant suitability: clinical

- Take a medical and social history, including drug compliance history.
- Perform a full physical examination, including breast and rectal examination.
- Assess for absolute contraindications, which include:
  - infection, severe cardiovascular, respiratory or liver disease, vasculitis or active glomerulonephritis, HIV and malignancy
  - current drug or alcohol abuse
- Assess for relative contraindications, which include:
  - age >70 years, uncorrected lower urinary tract disease, obesity, needle phobia and inability to provide informed consent
- Refer for psychiatric assessment in selected cases.
- Consider aetiology of renal failure and risk of recurrence of primary disease in renal graft.
- Refer to transplant physician and transplant surgeon for assessment.

PROTOCOL 2.8B

How to assess patients on dialysis for transplant suitability: laboratory investigations

- Full blood count, biochemistry and coagulation studies.
- HbA_{1C}, iron studies, vitamin B_{12}, folate, PTH, PSA (males).
- Hepatitis A, B and C serology, HIV, CMV, Epstein-Barr virus (EBV), human herpes virus 6.
- Urine culture and sensitivity.
- Tissue typing monthly.
- Mantoux testing in selected patients.
- Serology for *Strongyloides* in selected patients.

PROTOCOL 2.8C

How to assess patients on dialysis for transplant suitability: imaging

- CXR, ECG, abdominal ultrasound.
- Dipyridamole (Persantin) sestamibi scan in all patients with positive cardiovascular risk factor profile. Coronary angiography in patients with abnormal dipyridamole sestamibi scan results.
- Femoral duplex scan to evaluate iliac vessel patency for vascular anastomosis.
- Carotid duplex scan in patients at risk of cerebrovascular disease.
- Gastroscopy if there is a history of peptic ulcer disease.
- Colonoscopy for patients >50 years.
- Mammography for women >50 years.
- Cervical smear in all sexually active females >18 years.
- Pulmonary function tests in selected patients.
- Urodynamic testing in selected patients.
- Cystoscopy in selected patients.
- Dental review.

PROTOCOL 2.8D

How to assess the suitability of potential living kidney donors: clinical

- ABO blood group compatibility is the prerequisite for this process to proceed.
- Take a medical and social history.
- Perform a full physical examination, including breast and rectal examination.
- Absolute contraindications include:
  - infection, cardiovascular, respiratory, liver and renal disease, diabetes, malignancy, drug or alcohol abuse, pregnancy
  - age <18 years, inability to provide informed consent, coercion or financial inducement
- Relative contraindications include:
  - age >65 years, obesity, family history of diabetes, cigarette smoking, renal stones, uncorrected urinary tract abnormalities
- Refer for psychiatric assessment in selected cases.
- Refer to transplant physician and transplant surgeon for assessment.

PROTOCOL 2.8E

How to assess the suitability of potential living kidney donors: laboratory investigations

- ABO blood group compatibility is the prerequisite for this process to proceed.
- Full blood count, biochemistry and coagulation studies.
- 24-hour urinary protein and creatinine clearance.
- HbA_{1C}, iron studies, vitamin B_{12}, folate, PTH, PSA (males).
- Hepatitis A, B and C serology, HIV, CMV, EBV, human herpes virus 6.
- Urine culture and sensitivity.
- Oral GTT if there is a family history of diabetes.
- Pulmonary function tests in selected patients.
- If not checked in the previous 2 years, cervical smear in women between 18 and 69 years who have ever been sexually active.
- Tissue typing.

PROTOCOL 2.8F

How to assess the suitability of potential living kidney donors: imaging

- CXR, ECG, abdominal ultrasound.
- Formal measurement of GFR using technetium-99m DTPA.
- CT renal angiogram to evaluate renal vasculature.
- Dipyridamole (Persantin) sestamibi scan in all patients with positive cardiovascular risk factor profile.
- Mammography for women $>$50 years if not checked in the past 2 years.

References

1. Kidney Health Australia. Organ donation and transplantation in Australasia. Online. Available: <http://www.kidney.org.au>.
2. Gjertson DW, Cecka JM. Living unrelated donor kidney transplantation. Kidney Int 2000; 58(2): 491–9.
3. Liu PL, Gallery ED, Grigg R, et al. Renal function in unilateral nephrectomy subjects. Urol 1992;147(2): 337–9.

Chapter 2.9
Drug prescription to dialysis patients

HIGHLIGHTS

- When drugs are administered to dialysis-dependent patients, physicochemical drug characteristics and the mode of dialysis determine the extent to which drug pharmacokinetics are altered.
- When renal clearance contributes minimally to normal drug removal, CKD may not alter plasma clearance of the drug.
- For information on a particular drug, the prescriber should refer to the product information, referenced guides and Internet-based resources.
- Because patients are prescribed a large number of medications, the risks of non-compliance, drug mismanagement, drug interactions and side-effects are high.

Drug characteristics that influence dialysability

Drugs that are poorly removed by dialysis generally have:

- high lipid solubility
- a large volume of distribution (i.e. the drug is extensively distributed in tissues throughout the body with only a small amount in the blood)

- high plasma protein binding—only unbound drug is available to cross the dialysis membrane
- higher molecular weight—drugs with MWs >500–1000 Daltons are normally not eliminated by low-flux dialysers.

Drugs that are effectively removed by dialysis generally have:

- high water solubility
- low plasma protein binding
- lower molecular weight

Mode of dialysis and drug dialysability

The properties of the dialysis membrane, particularly pore size, influence drug removal.

- For HD, a higher dialyser surface area, higher blood and dialysate flow rates and longer time on dialysis may increase drug removal.
- High-flux dialysers and haemofilters used in continuous renal replacement therapies generally enhance drug removal of molecules up to 5000 Daltons (e.g. vancomycin is not removed by conventional HD but is removed by high-flux dialysers).

Removal of drugs by PD is generally modest.

- A higher PD exchange frequency may increase drug removal.
- The peritoneal membrane pores appear to be more permeable to larger molecular weight substances than are HD membranes.

General principles of drug prescription for patients on dialysis

- If possible, choose drugs that are cleared by non-renal routes or have a wide therapeutic margin.
- Avoid drugs with known toxic metabolites, such as pethidine.
- Avoid potential nephrotoxic drugs, such as aminoglycosides and NSAIDs, if patients have significant residual renal function.
- Monitor drug levels.

Guidelines to the prescription of new drugs for patients on dialysis are given in Protocol 2.9.

PROTOCOL 2.9

How to prescribe new drugs for patients on dialysis

- Review drug history (prescribed and patient initiated).
- Are there potential drug or complementary medicine interactions (e.g. decreased absorption of fluoroquinolones, iron salts or thyroxine when taken with phosphate binders)?
- Are contraindicated drugs being used (e.g. metformin)?
- Before increasing the dose of a drug (e.g. a phosphate binder) or adding new medications of the same class (e.g. an antihypertensive agent) assess the likelihood of patient compliance.
- Determine the appropriate dose and dose timing of the new drug. This may differ with the type of dialysis used.
- What loading dose is appropriate to the patient's age and weight? Although most drugs require a 'normal' loading dose, some do not (e.g. digoxin has a smaller volume of distribution in renal failure and requires a smaller loading dose).
- Does the maintenance dose require a reduced dose or dose frequency (e.g. digoxin, vancomycin, aminoglycosides)?
- If you are unfamiliar with a drug, confirm the dosage from more than one source (e.g. product information plus drug information reference).
- Monitor drug therapy for efficacy *and* toxicity using therapeutic drug monitoring when available. Assays of the free drug are best in cases of renal failure because altered levels of plasma proteins can make total drug levels difficult to interpret (e.g. phenytoin).

Adverse drug reactions

Adverse drug reactions are common because patients have:

- negligible renal clearance of drugs and their metabolites
- comorbidity
- a high risk of drug interactions or medication mismanagement because most patients are taking ≥8 regular medications

References

1. Ashley C, Currie A, eds. The renal drug handbook. 2nd edn. Oxford, UK: Radcliffe Medical Press, 2004.
2. Aronoff G, Aronoff GR, Berns JS, et al. Drug prescribing in renal failure: dosing guidelines for adults. 4th edn. Philadelphia, PA: American College of Physicians, 1999.
3. Therapeutic guidelines: antibiotic. Version 12. Melbourne: Therapeutic Guidelines Ltd, 2003.
4. Johnson CA, Simmons WD. Dialysis of drugs. 2004 edn. Ann Arbor, MI: Nephrology Pharmacy Associates Inc, 2004.
5. The Cochrane Renal Group. Online. Available: <http://www.cochrane-renal.org>.
6. The National Kidney Foundation. Online. Available: <http://www.kidney.org>.
7. International Society of Nephrology. Online. Available: <http://www.isn-online.org>.

Chapter 2.10
Nutrition

HIGHLIGHTS

- Ideally all patients starting dialysis should be seen by a dietitian.
- Prior to dialysis, in patients in CKD stages 4 and 5 a spontaneous reduction in protein intake is commonly observed. Nutritional support (e.g. moderate protein, high energy) may be required.
- Dietary restriction of high potassium foods is usually necessary for patients on HD but seldom for those on PD.
- Protein requirements for patients on HD are generally met but patients on PD may require protein supplements if intake is inadequate.
- Most patients will require phosphate binders except those using overnight or short daily HD.

Nutritional requirements for patients on dialysis

- Generally appetite and food intake improve when dialysis is commenced.
- Diets should be consistent with the healthy eating guidelines: 50% total energy from carbohydrates and 30–35% energy from fat (predominantly poly- and monounsaturated).
- Encourage patients to achieve a healthy body weight with a BMI of 20–25.

PROTOCOL 2.10

How to identify patients needing nutritional assessment

- Patients commencing dialysis.
- Patients changing dialysis modality.
- Patients with coexisting disease, such as diabetes.
- Patients with progressive reduction of body weight/BMI.
- Patients with anorexia, recurrent vomiting, diarrhoea or constipation.
- Presence of reduced serum albumin level.
- Presence of a 'low' serum creatinine level, reflecting a reduced muscle mass.
- Presence of a low serum cholesterol level, indicating inadequate energy intake.
- Persistent hyperkalaemia, hypokalaemia or hyperphosphataemia.

- Energy restriction may be required for overweight patients and PD patients (dextrose absorption from the dialysate can contribute to excessive weight gain).
- Other patients may need advice to increase their energy intake to avoid malnutrition.

Protein

- Patients on PD have high protein requirements as they lose 5–15 g of protein per day. Protein-rich supplements may be required if intake is inadequate.
- Protein requirements for patients on HD are generally met if a healthy diet is consumed.

Potassium

- Potassium dietary restriction is usually necessary for patients on HD but seldom for patients on PD. (Table 2.10A lists some high and low potassium foods.)
- Non-dietary causes of hyperkalaemia should be investigated (e.g. drugs including ACEI, constipation, inadequate dialysis) before restricting dietary intake.

Table 2.10A Examples of foods high and low in potassium[a]

High potassium foods	Low potassium foods
Banana, kiwi fruit, dried fruit, fruit juice, vegetable juice	Apple, pear, mandarin, most canned fruit (drained)
Potato, avocado, English spinach	Peas, beans, celery, onion
Soybeans	Rice, pasta
All Bran, muesli, Bran Flakes	Cornflakes, Rice Bubbles, puffed wheat, Weetbix
Nuts, chips, crisps, chocolate	

(*a*) High potassium foods should be avoided or limited; low potassium foods may need to be limited in some patients.

Phosphate

- Restricting phosphate intake is difficult as phosphate is widely distributed in foods, particularly in high protein foods (Table 2.10B).
- However, foods with a high phosphate content should probably be avoided.

Table 2.10B Examples of high phosphate foods to be avoided[a]

Offal (liver, kidney, brains)
Fish paste, meat paste, pâté, Vegemite, peanut butter
Nuts, seeds
Cola soft drinks, beer
Chocolate, carob and cocoa drinks

(*a*) Legumes are a rich source of phosphate but may be included in the diet in place of fish, chicken or dairy products.

Note: For further information, the average amount of phosphate per gram of protein in a variety of foods is available in the K/DOQI guidelines: K/DOQI. Kidney Disease Outcomes Quality Initiative. Am J Kidney Dis 2003; 42(4) (suppl 3): S69–70.

- Phosphate binders with meals or increased dialysis hours should be used in preference to phosphate restriction.

Sodium

- A 'no added salt' diet is encouraged to assist in fluid and BP control; avoid foods high in salt and do not add salt at the table.
- Note that many salt substitutes are not suitable as sodium is often replaced by potassium.

Fluids

- The recommended fluid intake will depend on the urine output, fluid balance and BP.
- Fluid from drinks, gravies, sauces, soups, jelly and yoghurt contributes to total intake.

Fats

- A low-saturated fat diet is recommended for all dialysis patients to help prevent cardiovascular disease.
- Poly- and monounsaturated fats should be used in preference to saturated fats.
- Aim for lean meats; low-fat dairy products and two fish meals a week.

Fibre

- A high fibre intake is important to prevent constipation, particularly in patients on PD, as well as helping in the management of hyperlipidaemia.
- Suitable high fibre foods include wholemeal breads and cereals, fruits and vegetables.
- A fibre supplement (e.g. Metamucil) may be required if dietary intake is inadequate.

Vitamins

- Vitamin supplementation may be required due to poor food intake, loss of vitamins across dialysis membranes and diminished absorption in dialysis patients.

- Vitamins B (usually as oral folate and multivitamins) and C supplementation are recommended.

Alcohol

- A small amount of alcohol can usually be included, provided it is kept within the fluid allowance.

Assessment of nutritional status

Clinical and biochemical indicators of poor nutrition are listed in Table 2.10C.

Table 2.10C Clinical and biochemical indicators of poor nutrition

Factor	Indication
Clinical	
Weight	≥5% unexplained weight loss in a month Gaining excessive dry body weight
BMI	<18.5
Fluid	Fluid gain >1 kg/day
Biochemical	
Serum albumin level	<32 g/L or progressively decreasing
Serum potassium level	>6.0 mmol/L after the longest HD interval
Serum phosphate level	>2 mmol/L routinely
Serum creatinine level	Progressively decreasing
Predialysis urea level	<20 mmol/L routinely
Total cholesterol level	>4.0 mmol/L[a]
HDL-C level	<1.0 mmol/L[a]
Triglyceride level	>2.0 mmol/L[a]
Blood glucose level	Persistently high

(*a*) At these levels the Heart Foundation recommends dietary intervention.

Methods used to assess nutritional status include:

- anthropometry: body weight, BMI (kg/m^2), weight history, body composition such as mid-arm circumference and triceps skin fold thickness
- nutritional tools: 'subjective global assessment' uses a systematic method of obtaining a nutritional history, particularly weight loss over the prior 6 months and recent trends over the prior 2 weeks; in conjunction with physical examination, the results are used to rate the patient's nutritional status; the reliability and validity of this nutritional tool are well established
- 24-hour recall or diet history with careful prompting; 3-day food diaries are also helpful

Nutritional support for malnourished patients

Dietary counselling

- Dietitians are located at most dialysis/renal units.
- Aim for a high energy/high protein diet.

Oral nutritional supplements (refer to dietitian)

- These are required if food intake remains poor.
- Supplements can be consumed as tolerated to increase energy and/or protein intake.
- Compliance can be an issue due to unpalatability and taste fatigue.

Enteral nutrition

- This is indicated when food and fluids are unable to be tolerated orally.
- Intradialytic parenteral nutrition or amino acid-containing dialysate may be used for the malnourished HD and PD patient, respectively.

Parenteral nutrition

- This is indicated when the gastrointestinal tract is non-functional.

References

1. Kidney Health Australia. CARI guidelines. Online. Available: <http://www.kidney.org.au/cari>.
2. K/DOQI. Clinical practice guidelines for nutrition in chronic renal failure. Am J Kidney Dis 2000; 35 (suppl 2): S1–S103.
3. National Heart Foundation of Australia and Cardiac Society of Australia and New Zealand. Lipid Management Guidelines 2001. Online. Available: <http://www.heartfoundation.com.au>.

Chapter 2.11
Routine surveillance of dialysis patients

HIGHLIGHTS

- Surveillance is the regular collection of data relevant to patient health and well-being.
- Access to these data should assist in patient care by medical, nursing, dietetic and pharmacy staff.

Table 2.11 is a guide to basic testing and will vary according to unit protocols and individual patient needs.

Infection surveillance is covered separately (see Table 2.6A).

Reference

1. CARI suggestions for clinical care. Online. Available: <http://www.kidney.org.au/cari>.

Table 2.11 A guide to routine surveillance

Tests	Monthly or second monthly	Third monthly	Yearly	Second yearly
Routine blood tests	Full biochemistry, including calcium by phosphate product and magnesium level; FBC	PTH, iron studies, fasting lipids	Vitamin B_{12} and folate levels; plasma aluminium level	
Blood tests for selected patients	Monthly tissue typing; coagulation studies; CRP	HbA_{1C}		
Dialysis adequacy		Urea reduction rate (*Kt/V*)	PET test[a]	
Vascular access		Recirculation measurement		
Disease complications			CXR, ECG	
Age-related screening				Pap smear (aged 18–69 years and ever sexually active); mammography (possibly aged >40 years and definitely aged 50–69 years)

(*a*) PET test at 6–12 monthly intervals unless changes to the dialysis prescription are made.

Part 3
Haemodialysis

The purpose of this part is to provide a basic framework to assist in the management of patients on HD.

Chapter 3.1
Haemodialysis physiology

HIGHLIGHTS

- HD is the term used to describe the process whereby the solute composition of blood is altered by its contact with another solution through a semipermeable membrane (see Fig. 3.1A).
- HD removes solutes from the blood by two processes:
  - diffusion (movement of a solute across the membrane down a concentration gradient)
  - convection ('solvent drag' or 'ultrafiltration'), whereby solutes in the blood are pushed through a membrane by a pressure gradient (analogous to filtration)
- Conventional HD relies predominantly on diffusion with a small component of convective transport.

- Sometimes blood purification is performed entirely by convection using a membrane of high permeability (HF). HF can also be combined with dialysis (HDF) but in this case most of the solute removal is by convection and not diffusion.

Determinants of solute transport during haemodialysis

The *extent* of removal for a given solute is determined by the *rate* of removal and the *duration* of treatment. Methods for assessing the extent of solute removal are discussed in Chapter 3.3.

Determinants of the rate of diffusive solute removal

- Blood flow rate through the dialyser
- Surface area of the dialysis membrane
- Permeability of the membrane (measured by the membrane efficiency or *KoA*; see Chapter 3.2)
- Dialysate flow rate
- Concentration gradient across membrane
- Molecular size of the solute—not only do large molecules move more slowly than small molecules but they may be too large to pass through the membrane pores

Determinants of the rate of convective solute transport

- Surface area and *water permeability* of the membrane (measured by the membrane 'flux' or *Kuf*; see Chapter 3.2)
- Membrane *porosity* (explained below)
- Blood flow rate through the dialyser
- Size of the solute
- Pressure differential across the membrane (transmembrane pressure, see below)

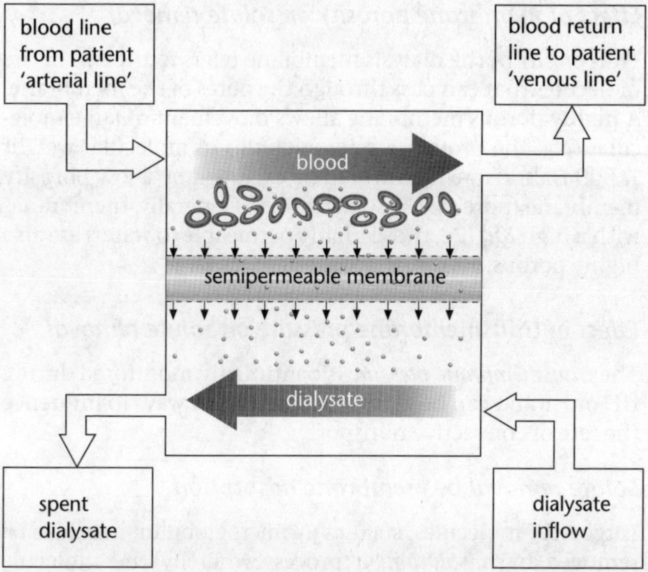

Fig. 3.1A Movement of solutes across a semipermeable membrane in HD. Blood and dialysate flow in opposite directions on either side of a semipermeable membrane (countercurrent flow), which maximises the removal of small-MW solutes from the blood. The semipermeable membrane only allows the passage of small-MW molecules and water.

Effect of membrane porosity on solute removal

The *porosity* of the dialysis membrane refers to the size of the molecules that can pass through the pores of the membrane. A highly porous membrane allows movement of large molecules (e.g. the protein β-microglobulin of molecular weight 11 800 Daltons), which would be excluded by a low porosity membrane (pore size < 500 Daltons). Generally, membranes with a high *Kuf* (i.e. those highly permeable to water) are also highly porous.

Effect of transmembrane pressure on solute removal

The *transmembrane pressure* is continually monitored during HD or HF and can be varied in a number of ways to influence the rate of convective transport.

Solute removal by membrane adsorption

Large-MW molecules, such as β_2-microglobulin, may also be removed by *adsorption*, a process whereby the molecule attaches to the dialysis membrane. The adsorptive capacity of a dialysis membrane is largely determined by its composition.

Methods of blood purification using diffusive and convective therapies

- Depending on the characteristics of the membrane used, several different methods can be used to remove fluid and uraemic toxins from the blood of an HD patient (Fig. 3.1B).
- The different properties of these methods are compared in Table 3.1.

Conventional haemodialysis

- In conventional HD, the blood and dialysate fluid compartments are separated by a semipermeable membrane (see Fig. 3.1A). These membranes can be made from various materials (see Chapter 3.2).

- The patient's blood flows in the opposite direction to the dialysate (countercurrent flow), which maximises diffusive solute transfer.

High-flux and high-efficiency haemodialysis

- Compared to conventional HD, high-flux membranes rely, to a larger degree, on convective solute transport, which makes them more effective at removing higher molecular weight solutes, such as β_2-microglobulin.
- Although observational studies have demonstrated improvements in BP control, fluid status, anaemia, LVH, phosphate control and neuropathic symptoms using high-flux therapies, there are no randomised controlled studies evaluating the effectiveness of these modalities on long-term outcomes, such as mortality, cardiovascular events or bone disease.
- Most high-flux membranes in use can also be classified as 'high efficiency' because of their large surface area, making them also more efficient at removing low-MW solutes. Although there was initially some interest in using these membranes to shorten dialysis treatment times, this approach is not recommended (see Chapter 3.3).

Haemofiltration and haemodiafiltration

- As mentioned above, HF relies solely on convective solute transport to remove uraemic toxins from the blood. HF is most commonly used in patients receiving continuous treatment in the intensive care setting. Replacement fluid is infused directly into the patient's blood, usually using a preprepared sterile solution.
- HDF combines HF with HD to introduce a component of diffusive transport. In the chronic setting, this therapy is most effective at removing high-MW solutes but is

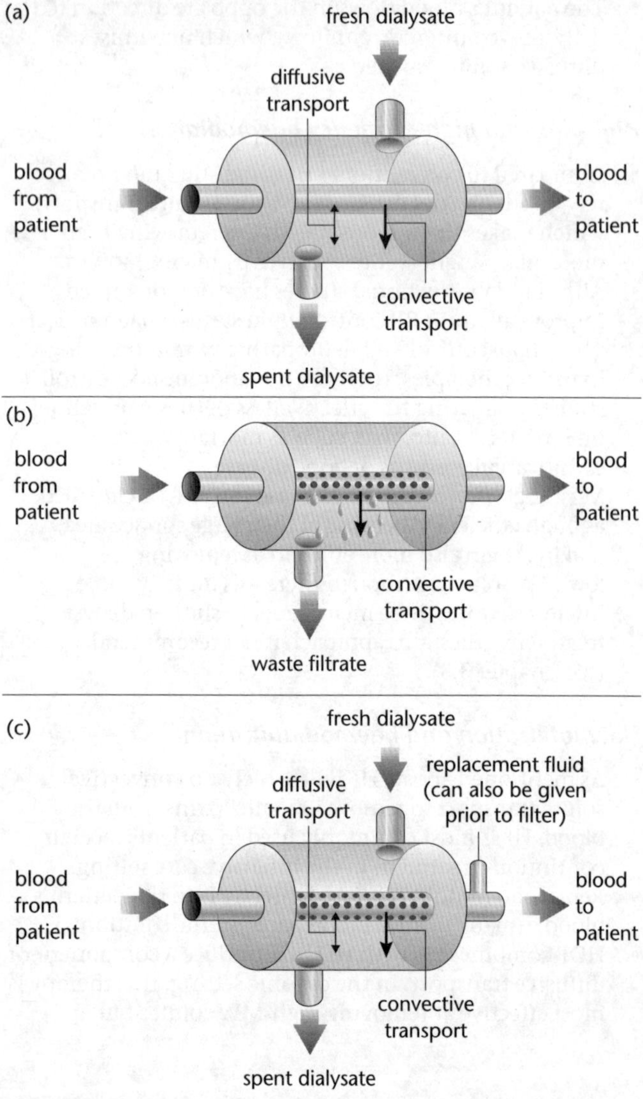

Fig. 3.1B Comparison among (a) haemodialysis, (b) haemofiltration and (c) haemodiafiltration

Table 3.1 Comparison among haemodialysis, haemofiltration and haemodiafiltration

	Treatment type			
	Conventional HD	**High-flux–high-efficiency HD**	**HF**	**HDF**
Diffusive transport	Yes	Yes	No	Yes
Convective transport	Minimal	Moderate with high blood flow rates	Extensive	Extensive
Removal of low-MW solute	Good	Excellent	Poor	Excellent
Removal of high-MW solute	Poor	Moderate	Good	Excellent
Application	Chronic dialysis	Chronic dialysis	Continuous dialysis in critically ill patients	• Continuous dialysis in critically ill patients • Chronic dialysis in selected patients

considerably more expensive and requires ultra-pure dialysis water (since the fluid removed by filtration is replaced into the blood side of the dialysis circuit).

- Because of its expense, HDF is generally reserved for patients receiving continuous treatment in the intensive care setting or chronic dialysis patients who have complications possibly related to the accumulation of high-MW uraemic toxins, including β_2-microglobulin-associated amyloidosis and uraemic neuropathy.

Chapter 3.2
Haemodialysis equipment

HIGHLIGHTS

The essential components for HD (shown in Fig. 3.2) include the:
- dialyser (artificial kidney)
- dialysis solution (dialysis water and concentrates)
- dialysis tubing
- HD machine

Dialyser

The dialyser (the 'artificial kidney') is the apparatus in which the dialysate fluid and blood are in contact across a semi-permeable membrane.

Ports

There are four ports attached—two for the dialysate and two for the blood. Blood flowing from the patient to the dialysis machine is carried along the 'arterial' limb of the circuit. Conversely, the 'venous' limb carries purified blood from the dialysis machine back to the patient.

Dialysis membrane

The semipermeable membrane is suspended within the dialyser. These can be arranged in two different configurations:

- *hollow fibre* (or capillary) *configuration*—contains thousands of hollow fibres through which the blood flows, while the dialysate flows outside the fibres within the dialyser; most dialysers used in adults are of this type

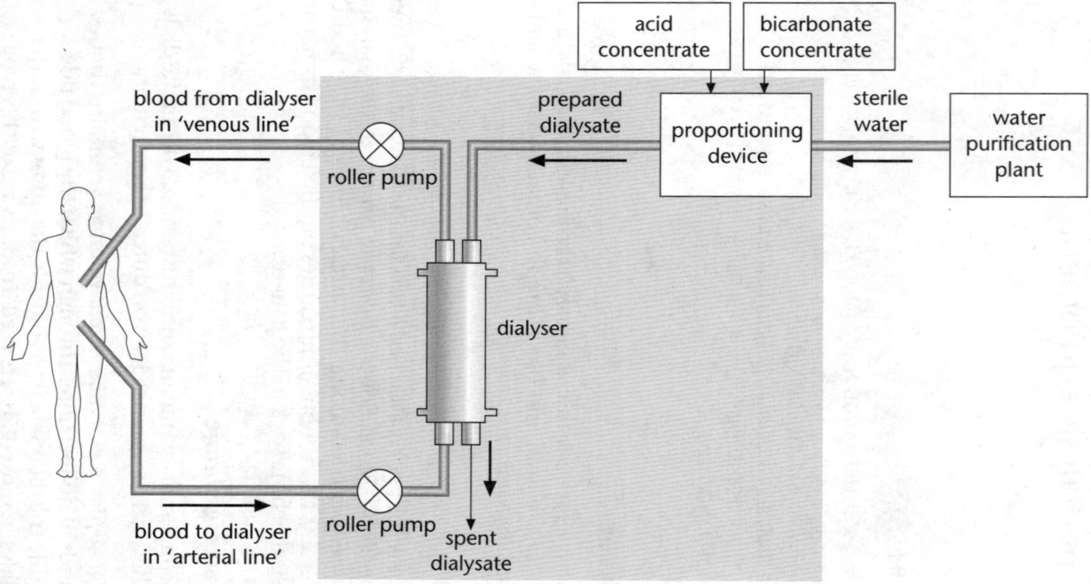

Fig. 3.2 Components of a typical haemodialysis machine

- *parallel plate configuration*—contains flat sheets of membrane material stacked on top of each other, arranged so that blood and dialysate fluid flow in alternate spaces; there is low flow resistance, making them particularly useful for 'single needle' dialysis in children

Membrane composition

- *Cellulose* (cuprophan or cuprophane)—made from cotton (cellulose polymer). Less 'biocompatible' than other membrane types due to activation of inflammatory pathways (see Chapter 3.7).
- *Substituted cellulose*—cellulose polymer in which the free hydroxyl groups are substituted.
- *Cellulosynthetic*—synthetic material added to liquefied cellulose. Examples include Cellosyn and Haemophan.
- *Synthetic*—not cellulose-based. Examples include polysulfone, polyacrylonitrile and polyamide.

Dialyser specifications

- *Surface area*—usually 0.7–2.4 m^2. The larger the surface area, the higher the clearance (and priming volume).
- *Dialyser efficiency*—measured by the *mass transfer coefficient* (*KoA*). The *KoA* is a function of the permeability of the membrane and its surface area. A high-efficiency dialyser is one with a *KoA* of greater than 600–700 mL/min (usually achieved by a larger membrane surface area).
- *Ultrafiltration coefficient* (*Kuf*)—this determines the degree of convective transport (flux) achieved by a particular dialyser. High-flux membranes have large pore sizes and allow greater removal of high-MW solutes.

Dialyser solution (dialysate)

- Dialysate is made by mixing treated water with preprepared ionic concentrates using a proportioning device within the HD machine.

- The typical composition of a dialysate solution is shown in Table 3.2.

Water for haemodialysis

- In dialysis centres, water for dialysis is generated in a central water treatment plant (a scaled-down version is used for patients on home HD).
- Dialysis water is purified in a number of steps, including softening (removal of calcium and magnesium), passage through a carbon filter (removing impurities such as chloramine), removal of small-MW solutes (either by reverse osmosis or deionisers) and sterilisation (by passage through a bacterial filter).
- Impurities that should be removed include:
  - aluminium—may cause 'dialysis dementia', bone disease and anaemia
  - chloramine—may cause haemolysis
  - bacteria
  - endotoxin—usually filtered out by the dialyser membrane, but not by the high-flux membranes; may cause pyrogenic reactions on dialysis
  - copper—may cause haemolysis
  - fluoride—may cause bone disease (chronic) or death (acute)

Table 3.2 Typical composition of a bicarbonate-based dialysate

pH	7.1–7.3
Sodium	135–145 mM
Potassium	2–4 mM
Chloride	90–120 mM
Calcium	0.9–1.7 mM
Magnesium	0.3–0.8 mM
Bicarbonate	27–40 mM
Dextrose	5–10 mM

- Recommendations exist for the minimum levels of these contaminants in water for dialysis (refer to the AAMI guidelines in the reference list for more information).
- Regular testing of water quality is recommended.

Dialysate concentrates

In bicarbonate-buffered dialysis, two separate concentrates are mixed with the treated water.

- The *acid* concentrate contains most of the important ionic constituents of blood (i.e. sodium, potassium, chloride, calcium, magnesium and dextrose).
- *Bicarbonate* is added from a separate concentrate reservoir—it would form insoluble calcium carbonate if stored in the same concentrate.

Dialysis tubing

- Disposable tubing connects the dialyser to the patient.
- Low-priming volume lines are used for dialysing children in the paediatric setting.

Haemodialysis machine

Standard components

- Roller pumps for dialysate and blood paths.
- An infusion pump for heparin.
- Pressure monitors on the 'arterial' and 'venous' limbs of the blood circuit.
- Air trap/air detector and blood leak detector.
- Dialysate heater and dialysate temperature monitor.
- Conductivity monitor (to monitor dialysate sodium).

Optional devices

- Dialysate urea sensor (allows the continuous calculation of urea removal as an index of dialysis dose).
- Blood volume sensor (monitors change in haematocrit and volume status during dialysis).

Reference

1. Association for the Advancement of Medical Instrumentation (AAMI) standard for dialysis water quality. 3rd edn. 1998. Online. Available: <http://www.aami.org>.

Chapter 3.3
Haemodialysis adequacy and prescription

HIGHLIGHTS

- Patients on maintenance HD should have the effectiveness (adequacy) of their dialysis treatment assessed on a regular basis.
- The effectiveness of dialysis should be assessed using a combination of indicators and *not* depend entirely upon the achievement of a target level of urea clearance (see below).

What is 'adequate' haemodialysis?

Adequate HD is a treatment regimen that is acceptable to the patient and dialysis provider in terms of:

- control of uraemic symptoms and patient well-being
- BP control
- biochemical markers of uraemia and nutrition (see Chapter 2.10)
- achieved dialysis dose based on small solute clearance
- inconvenience and cost

Although the effectiveness of dialysis is partly dependent upon the amount of dialysis the patient receives, the adequacy should *not* be entirely based upon achievement of a particular dose 'target'.

Recommendations for adequacy assessment

The adequacy of dialysis should be assessed using a combination of indicators (see Protocol 3.3A).

The recommended frequency of assessment is:

- 3-monthly in stable patients
- monthly in unstable patients or after a change in dialysis prescription

PROTOCOL 3.3A

How to assess the adequacy of delivered haemodialysis

Subjective assessment

- How well are the symptoms of uraemia controlled in the patient (e.g. appetite, nausea, tiredness, itch)?

Objective assessment

- Volume/BP control.
- Is the patient still acidotic (as indicated by a low mid-week predialysis bicarbonate level)?
- How well are blood phosphate levels controlled (see Chapter 2.4)?
- Is the serum albumin level normal (an indicator of nutrition strongly associated with survival; see Chapter 2.10)?

Dialysis dose assessment

- Is the desired level of urea removal being met?

Approach to the patient with suspected inadequate dialysis

Protocol 3.3B gives an algorithm for the assessment of suspected inadequate HD.

PROTOCOL 3.3B

How to assess suspected inadequate haemodialysis

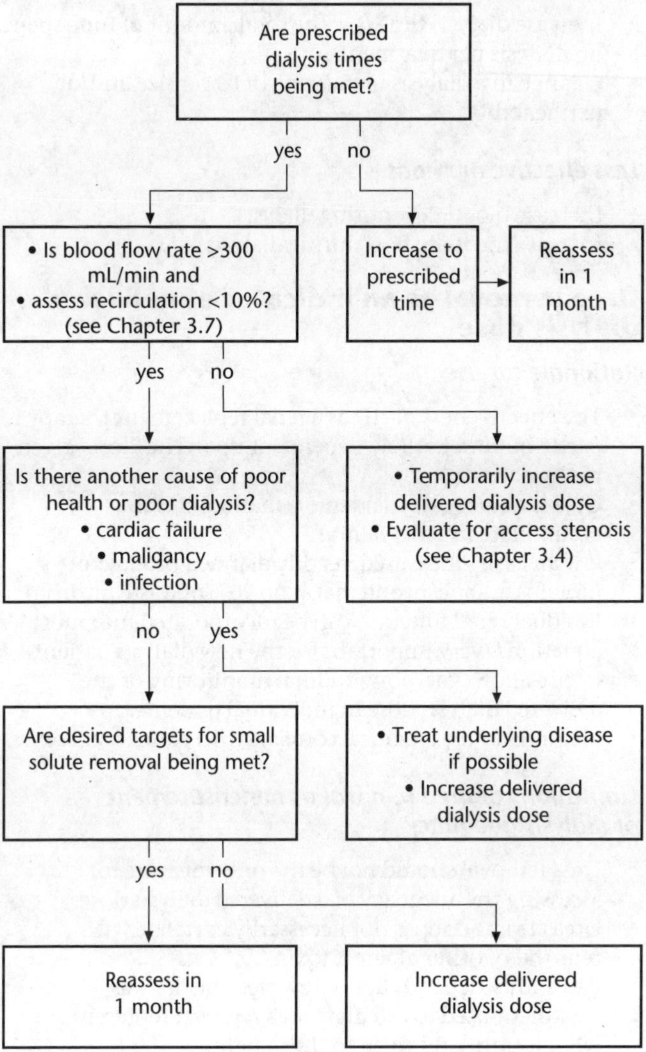

Methods to increase delivered dialysis dose

Most effective methods

- Increase dialysis time (i.e. the total amount of time spent on dialysis per treatment).
- Change to a dialysis membrane of larger size and/or permeability.

Less effective methods

- Increase blood flow during dialysis.
- Increase dialysate flow during dialysis.

Urea removal as an indicator of delivered dialysis dose

Rationale for use

- The effectiveness of HD as a renal replacement therapy is partly but not entirely dependent upon the 'dose' given to the patient.
- Urea is a surrogate marker for small solutes that accumulate in renal failure.
- It is an easily measured, readily dialysed product of protein catabolism and makes up 90% of waste nitrogen.
- Residual renal function can be incorporated into most equations (very important for the new dialysis patient).
- Its use allows easy longitudinal monitoring of the delivered dialysis dose in individual patients.
- Measures of urea removal correlate with patient outcomes.

Limitations of urea removal as a measurement of dialysis adequacy

- Urea removal should not be the only measure for assessing the adequacy of a delivered dialysis dose.
- Urea clearance does not necessarily correlate with removal of other uraemic toxins.
- Its validity depends upon how well the dialysis treatment used to calculate urea removal represents what is normally given to the patient.

- The accuracy of measurement depends on the precision and timing of post-dialysis urea levels (which 'rebound' after dialysis; see Protocol 3.3C).

PROTOCOL 3.3C

How to optimise the measurement of post-dialysis urea levels

Rationale

- Urea levels 'rebound' after dialysis due to equilibration of urea levels within blood and the extracellular fluid.
- If this is not taken into consideration, post-dialysis urea levels will be artefactually low and measures of urea removal will be artificially high.

Suggested method

- Set UF rate to zero.
- Slow blood pump to 50 mL/min for 10 seconds.
- Stop blood pump.
- Take blood from 'arterial' line after 1 minute.

Methods of assessing urea clearance

See Table 3.3A and Protocol 3.3D on how to estimate urea clearance using mathematical formulae.

- **Proportional reduction of urea** (urea reduction ratio or *URR*)—simplest and easiest to calculate.
- **Estimated** urea clearance *(Kt/V)* using a mathematical formula—*Kt*/*V* is an estimate of urea removal based on the dialyser efficiency (*K*) multiplied by the time of dialysis (*t*) and divided by an estimation of body water content (*V*).
- **Calculated** *Kt*/*V* using urea kinetic modelling—three measures of urea are required. This method also enables calculation of the protein catabolic rate (*nPCR*—an indicator of protein intake in the steady state).

- **Calculated** *Kt/V* by the measurement of urea removal in the dialysate. Some dialysis machines are able to estimate urea removal by continuous estimation of the urea concentration in the fresh and spent dialysate.

Table 3.3A Characteristics of methods used for the assessment of urea removal

Method	Advantages	Disadvantages
Proportional reduction of urea (*URR*)	Simple	• Does not allow estimation of *nPCR*
Estimated equilibrated *Kt/V* using formulae	Moderately easy	• Does not allow estimation of *nPCR*
Kt/V using urea kinetic modelling (*UKM*)	Enables estimation of *nPCR*	• More complex • Requirement for three samples • Inaccuracy in calculation of *V* using formulae
Kt/V using the dialysate collection method	Most accurate	• Specialised equipment necessary
Kt/V using 'online' machine measurement of dialysate urea	Accurate	• Specialised dialysis equipment necessary • Ability to predict outcome not known

Suggested minimal levels of urea removal and/or protein catabolic rate (when measured)

- CARI guidelines — *Kt/V* >1.2, *URR* >65%
- K/DOQI guidelines — *Kt/V* >1.2, *URR* >65%, *nPCR* >1 g/kg/d
- UK Renal Association guidelines — *Kt/V* >1.2, *URR* >65%, *nPCR* >1 g/kg/d

PROTOCOL 3.3D

How to estimate urea clearance using mathematical formulae

Urea reduction ratio (URR)

$$URR = \frac{\text{predialysis urea} - \text{post-dialysis urea}}{\text{predialysis urea}} \times 100$$

Estimated single pool Kt/V (Daugirdas equation)

$$Kt/V = -\log_e\left[\frac{\text{post-dialysis urea}}{\text{predialysis urea}} - 0.08 \times t\right] + \left[4 - 3.5 \times \frac{\text{post-dialysis urea}}{\text{predialysis urea}}\right] \times \left[\frac{\text{predialysis weight} - \text{post-dialysis weight}}{\text{predialysis weight}}\right]$$

where t = dialysis time (hours).

Prescribing chronic haemodialysis

Basic dialysis prescription

'Prescribing' HD for patients in a maintenance dialysis program takes into account more than just the desired HD dose. Elements of the basic HD prescription include:

- desired minimal dialysis duration (not <4 hours at each treatment)
- desired frequency of dialysis (not <3 sessions/week unless residual GFR >10 mL/min)
- desired dry weight of the patient (enabling the desired volume of fluid removal to be calculated from the predialysis weight) (see Protocol 3.3E)
- intended level of urea removal
- anticoagulation protocol to be used (see Chapter 3.5)
- membrane type
- membrane size (surface area)—can be decided empirically (i.e. a 'big' membrane for a big patient and a 'small' membrane for a small patient) or based upon the estimated urea removal (see Protocol 3.3F)

PROTOCOL 3.3E

How to assess the dry weight of a haemodialysis patient

- The dry weight (also referred to as the IBWt) of an HD patient is the weight at which there is no excess fluid (either intravascular or extravascular).
- The estimated dry weight is used to set the amount of fluid removal at each HD session. If, for example, the estimated dry weight of the patient is 79 kg and his or her predialysis weight is 82 kg, then the targeted fluid removal for that patient during haemodialysis will be 3 kg.
- Achieving the dry weight during a dialysis session is dependent upon the patient's cardiovascular stability (see also Chapter 3.7).
- If the estimated dry weight is too *low*, then the patient may develop symptoms and signs of volume depletion during or following dialysis (cramps, light-headedness, dizziness, blackouts, low BP).
- If the estimated dry weight is too *high*, then the patient may develop symptoms and signs of volume overload before the next dialysis session (breathlessness, oedema, HT, pulmonary oedema).
- In practice, initial determination of the dry weight of the patient is 'trial and error' using the signs above and frequent reassessment.
- Factors to consider when estimating the dry weight of a dialysis patient include:
    - What is the timing of estimation in relation to their next dialysis session?
    - Is there any evidence of dependent oedema (as a rough rule of thumb, if this is present then the dry weight is at least 2–3 kg lower than the current weight)?
    - Is the JVP elevated (in the absence of right heart failure an elevated JVP would suggest that the dry weight is at least 2–3 kg lower than the current weight)?
    - Is the patient hypertensive (most HT in dialysis patients is fluid related)?

PROTOCOL 3.3F

How to estimate urea removal using the dialyser specifications

- The urea clearance (*Kt/V*) for an individual patient can be estimated using:
  - *K*, calculated from the urea clearance of the dialyser in use (*KoA*) and the blood flow
  - *t*, the dialysis time in minutes
  - *V*, calculated using anthropometric formulae
- The accuracy of this assessment should be checked within 1 week after prescription using one of the methods described above.
- In theory, this method could be used to estimate the dialysis duration (*t*) or membrane size (*KoA*) for a given patient/dialyser combination.
- Because of the unresolved question as to whether dialysis time is an independent determinant of dialysis outcome, using this approach to prescribe dialysis times shorter than 4 hours is not recommended.

Advanced dialysis prescription

- Components of the basic dialysis prescription can be further modified in specific circumstances.
- As discussed in Chapter 3.2, dialysate for HD machines is usually generated by a proportioning device within the dialysis machine that mixes dialysis water with concentrates of acid and bicarbonate.
- Because of this, the individual components of the dialysate can be modified under specific circumstances, as detailed in Table 3.3B (unless the dialysis unit prepares its dialysate using a central proportioning system).

Table 3.3B Elements of the advanced dialysis prescription

Parameter	Conventional level	Modification(s)	Effect/comment
Dialysate flow	500–800 mL/min	Increase	• Small effect on clearance • Increased water/concentrate consumption
Dialysate temperature	37.5–38°C	Reduce to 34–36°C	• Increased haemodynamic stability • May cause mild symptomatic hypothermia
Dialysate sodium	137–140 mmol/L	Increase to 140–145 mmol/L	• May increase haemodynamic stability and reduce cramping during HD • Risk of increased thirst and interdialytic weight gain
		Reduce to 130–135 mmol/L	• Consider if the patient has a predialysis sodium level <125 mmol/L • May reduce interdialytic weight gain • Risk of intradialytic hypotension
		'Ramped' profile (high at start, low towards end)	• May increase haemodynamic stability and reduce cramping during HD

Parameter	Conventional level	Modification(s)	Effect/comment
			• Risk of increased thirst and interdialytic weight gain
Dialysate calcium	1.5 mmol/L	Reduce to 0.9–1.3 mmol/L	• Consider if severe hypercalcaemia • Occasionally used to allow increased oral calcium-based phosphate binders or in adynamic bone disease to stimulate PTH and bone turnover
		Increase to 1.7–2.0 mmol/L	• Consider if severe symptomatic hypocalcaemia (e.g. post parathyroidectomy)
Dialysate potassium	2 mmol/L	Increase to 3 mmol/L	• May reduce risk of arrhythmia in patients on digoxin or at risk of arrhythmia • Risk of hyperkalaemia
UF rate	Constant rate	Initial period of intense and/or 'isolated' UF followed by slower UF	• Useful in symptomatic fluid overload • May provide greater haemodynamic stability

References

1. CARI guidelines 2002. Online. Available:<http://www. kidney.org.au/cari/index.htm>.
2. Renal Association. Treatment of adults and children with renal failure: standards and audit measures. 3rd edn. London: Royal College of Physicians of London and the Renal Association, 2002. Online. Available:<http://www.renal.org/Standards>.
3. K/DOQI guidelines 2000. Online. Available:<http://www. kidney.org/professionals/kdoqi>.

Chapter 3.4
Vascular access for haemodialysis

HIGHLIGHTS

- Patients on maintenance HD treatment require stable access to the circulation that is capable of supporting blood flows of 250–400 mL/min.
- One of the major problems in HD patients is obtaining access to the patient's circulation that is reliable, safe and repeatable.
- The three basic methods for accessing a patient's circulation for HD are:
  - indwelling central venous catheter (CVC)
  - native AV fistula
  - synthetic AV fistula (AV graft)

  Each one of these methods has its own indications, advantages, disadvantages and complications.

Haemodialysis catheters

HD catheters are IV catheters with the tip lying in the right atrium, the great veins of the chest or occasionally the great veins of the abdomen. (Fig. 3.4A).

There are two types of HD catheters: *tunnelled* and *non-tunnelled*. They have the advantage of providing immediate access to the circulation if urgent dialysis is required.

CVCs for HD differ in a number of ways from other CVCs (Fig. 3.4B):

- a *double lumen* for circulation of blood between the patient and dialyser; the arterial limb (which takes blood to the dialyser) is shorter than the venous limb to prevent recirculation (see Chapter 3.7)

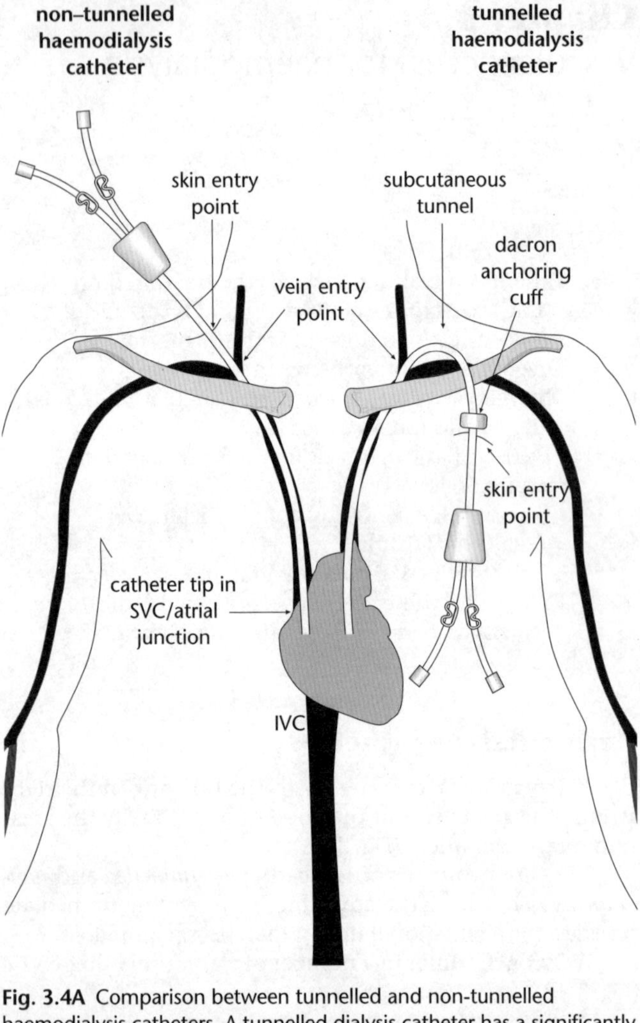

Fig. 3.4A Comparison between tunnelled and non-tunnelled haemodialysis catheters. A tunnelled dialysis catheter has a significantly longer subcutaneous tunnel. For the tunnelled catheter, the tunnel and the physical barrier formed by tissue ingrowth around the dacron cuff substantially reduce the rate of systemic infection. For both catheter types the ideal placement of the tip is in the lower SVC or SVC/right atrial junction

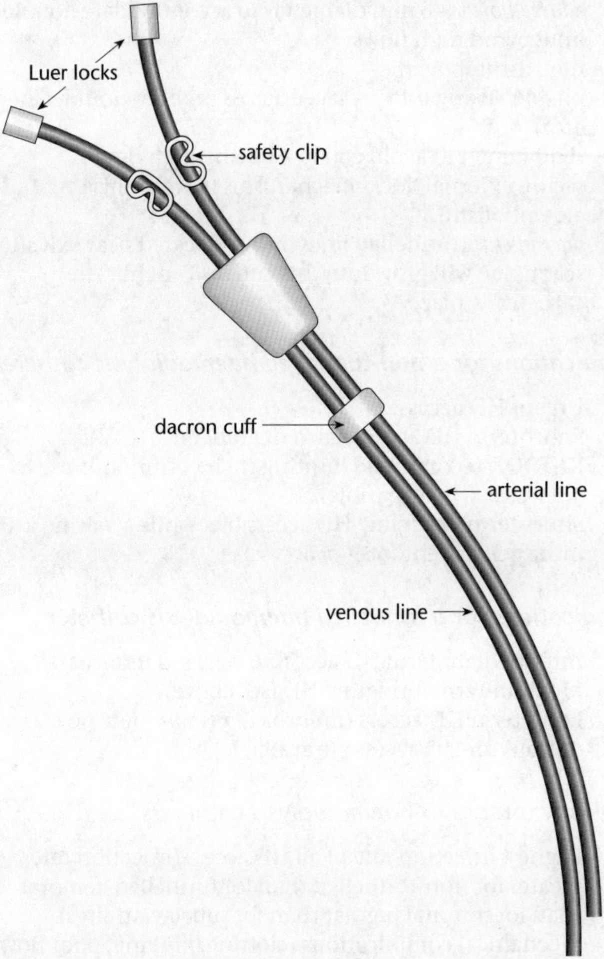

Fig. 3.4B Features of a tunnelled double lumen haemodialysis catheter. The arterial lumen and venous lumen are separated to prevent recirculation of blood. Non-tunnelled lines are similar in layout except that they are shorter and do not include a subcutaneous dacron cuff.

- a *large bore* (4–6 mm diameter) to accommodate double lumen and high flows
- non-thrombogenic
- pliable/strong (often placed across mobile shoulder/neck area)
- rigid lumen to avoid collapse during high flow
- secure external 'lock' mechanisms (to minimise venous air embolism)
- *dacron cuff* (tunnelled lines only)—located near exit site; scar tissue will grow into this cuff and anchor the catheter in place

Indications for a non-tunnelled haemodialysis catheter

- Urgent HD access.
- Short-term HD access (current guidelines (CARI, K/DOQI) recommend limiting their continuous use to <3 weeks where feasible).
- Short-term 'bridging' HD access (i.e. while awaiting a more permanent form of access).

Indications for a tunnelled haemodialysis catheter

- Intermediate-term HD access (6 weeks–3 months).
- Medium-term 'bridging' HD (see above).
- Long-term HD access (months to years, where no other options for HD access are available).

Disadvantages of haemodialysis catheters

- Highest infection rate of all HD access (infection rates greater for non-tunnelled than for tunnelled, femoral than for internal jugular than for subclavian sites).
- Mechanical complications (clotting, kinking, poor flows) may require replacement or 'rewiring' (replacement of the catheter over a guide wire).
- Non-tunnelled lines are often uncomfortable for and unattractive to the patient.
- Inability of the patient to ambulate with a non-tunnelled femoral line.

Method of placement

Tunnelled HD catheters are normally placed using fluoroscopy in either the operating theatre or radiology department and are not discussed here. See Protocol 3.4A for placement guidelines.

PROTOCOL 3.4A

How to insert non-tunnelled haemodialysis catheters

Choice of site (in order of preference)

- Right internal jugular vein (short, straight path to SVC/RA junction).
- Left internal jugular vein.
- Right or left femoral vein (highest infection rate).
- Right or left subclavian vein (greater risk of pneumothorax, risk of central venous stenosis which may make future AV fistula formation difficult).

Technique tips

- Strict sterile technique.
- Patient head down to distend vein and avoid air embolism.
- Ultrasound-guided puncture for jugular and femoral vein catheters.
- Secure fixation of catheter with monofilament nylon suture.
- X-ray after procedure to confirm catheter tip position at SVC/RA junction and to assess for pneumothorax/ haemothorax (not necessary for femoral catheters).

Complications of haemodialysis catheters

Procedural complications

- Inability to access the central veins (usually due to central vein occlusion or stenosis following repeated central venous catheterisation).
- Arterial injury (carotid, subclavian, femoral artery).
- Pneumothorax, haemothorax.

- Venous injury.
- Air embolism.
- Contrast allergy (tunnelled).

Short-term complications

- Exit site infections—Protocol 3.4B gives the assessment and management of HD catheter exit site infections.
- Systemic infection (particularly with non-tunnelled catheters)—Protocol 3.4C gives the assessment and management of HD catheter-related sepsis.
- Line occlusion (kink, in-line thrombus, position)—Protocol 3.4D shows how to manage suspected blockage of an HD catheter.
- Dislodgment (for a tunnelled line the dacron cuff needs at least 2 weeks to anchor).
- Air embolism.
- DVT of central veins.

Long-term complications

- As for short-term complications.
- Fibrous stricture stenosis/occlusion of central veins—this may impair the ability to form AV access in the ipsilateral arm.
- Catheter fatigue/fracture (leak, air embolism, catheter embolism).

PROTOCOL 3.4B

How to assess and manage catheter exit-site infections

- Defined by the presence of cellulitis or discharge around the skin insertion site.
- A risk factor for the development of catheter-related sepsis (see Protocol 3.4C).

Assessment and management

- Take a swab for microbial culture and sensitivity.
- Treat as for suspected catheter-related sepsis (see below) in the presence of fevers or rigors.
- Treat with appropriate antibiotics:
  - IV gentamicin 80 mg (given at the end of a dialysis treatment)
  - Oral anti-staphylococcal treatment (e.g. dicloxacillin 500 mg qid, cephalexin 500 mg qid or clindamycin 300 mg tds)
  - IV vancomycin 1 g if the patient is known to be colonised with MRSA
- Review antibiotics depending on the sensitivities of the microbial culture.
- Consider catheter removal if alternative access is available or refractory to treatment.

PROTOCOL 3.4C

How to assess and manage haemodialysis catheter-related sepsis

- Associated with high morbidity and mortality (up to 10%).
- Should be suspected in any patient with fever and an HD catheter (regardless of the duration of catheterisation).

Prevention
- Minimise the use of non-tunnelled dialysis catheters to less than 3 weeks wherever feasible (CARI/K/DOQI guidelines).
- Prompt treatment of exit site infections.
- Keep the catheter insertion site clean and dry.
- Mupirocin ointment to catheter insertion site.
- Consider the use of antibiotic-impregnated catheters.

Assessment of suspected cases
- Evaluate for other potential causes of fever.
- Obtain cultures of blood.

Management
- Treat empirically with broad-spectrum parenteral antibiotics (e.g. IV gentamicin 80 mg given at the end of dialysis with systemic cephazolin or dicloxacillin; IV vancomycin should be used if the patient is known to be colonised with MRSA).
- All non-tunnelled lines should be removed in all cases of suspected access-related sepsis and the catheter tip sent for bacterial culture.
- Patients with a cuffed and tunnelled dialysis catheter who are febrile may be given a 24–48 hour trial of antibiotics, failing which the catheter should be removed.
- Signs of tunnel infection, persistent sepsis or fungal septicaemia are additional indications for removal of a cuffed and tunnelled HD catheter.

Follow-up
- If the patient is found to be bacteraemic, a minimum of 7 days of antibiotic therapy is indicated. A 14-day duration of therapy is recommended for patients with *Staphylococcus aureus* bacteraemia.

PROTOCOL 3.4D

How to manage suspected blockage of a haemodialysis catheter

- Usually indicated by a failure to aspirate blood or attain reasonable flow rates during HD. May be caused by either:
  - apposition of catheter side–holes against vessel wall, or
  - catheter thrombosis (either intraluminal or due to a fibrin 'sheath' around the catheter tip)

Assessment and management

- Confirm catheter tip position radiographically.
- Gently aspirate lumen using a 2 mL syringe (may dislodge an intraluminal thrombus).
- Non-tunnelled lines:
  - Sometimes catheter function can be improved by rotation of the catheter by 180° within the securing device.
  - Exchange over a guide wire or resite in an alternative location.
- Tunnelled lines:
  - Consider intraluminal thrombolysis (see Protocol 3.4E).

Prevention

- Lock catheter lumens with a volume equal to that of the internal lumen (written on the catheter) using a concentrated solution of heparin (5000 U/mL).

PROTOCOL 3.4E

How to use thrombolytic agents for the treatment of tunnelled haemodialysis catheter thrombosis

Overview

- Both urokinase and recombinant tissue plasminogen activator (alteplase) can be used for thrombolysis.
- This protocol should be used for tunnelled HD catheters in which incorrect tip placement has been excluded. Non-tunnelled catheters should be changed over a guide wire if thrombotic occlusion is suspected.

Contraindications

- Known hypersensitivity to either urokinase or alteplase.
- Active bleeding, thrombocytopenia, haemostatic defects or those at high risk of embolic complications.
- Recent (within 48 hours) surgery, obstetric delivery, percutaneous biopsy of viscera or deep tissues, or puncture of non-compressible vessels.
- Suspected catheter infection (use of thrombolytics may release a localised infection into the systemic circulation).

Procedure

- Obtain consent from the patient
- Dilute appropriately:
    - alteplase 2 mg/mL—if frozen aliquots are used, thaw to room temperature by slow inversion (do not shake)
    - urokinase 5000 U/mL
- Administer thrombolytic solution into the occluded limb of the catheter. The volume given should be either 2 mL or the volume of the occluded limb (whichever is greater).
- After 30 minutes of dwell time, assess catheter function by attempting to aspirate blood. If catheter is functional, aspirate and discard the first 5 mL of blood obtained.
- If the catheter cannot be aspirated, repeat after 120 minutes. If the catheter can be aspirated, discard the first 5 mL as above.
- If the catheter still cannot be aspirated, either the dose can be repeated or alternative access sought.

Arteriovenous fistulae

- These are generally regarded as the access of choice for HD.
- AV fistulae are created by the surgical connection of a peripheral vein to a peripheral artery. The peripheral vein, under arterial pressure, then becomes much enlarged, making puncture easy and reliable, and develops a large blood flow, allowing for successful HD without recirculation.
- AV fistulae are usually named according to the vein and artery used in creation (e.g. a radiocephalic fistula is formed using the radial artery and cephalic vein at the wrist).
- The commonest types in use are the radiocephalic fistula (Fig. 3.4C) and the brachiocephalic fistula (Fig. 3.4D).

Advantages and disadvantages of arteriovenous fistulae

Table 3.4 gives a comparison between AV fistulae and AV grafts.

Creation of an arteriovenous fistula

Preoperative

- Find a suitable vein (by physical examination in thin patients with prominent veins or by ultrasound).
- Find a suitable adjacent artery. Arterial insufficiency is a common problem in diabetic patients, leading to difficulties with creation, poor maturation and arterial steal syndrome (see below).
- Instruct the patient to protect the designated arm/vein from venipuncture, IV lines or any other injury or interference (the forearm becomes the patient's kidney).

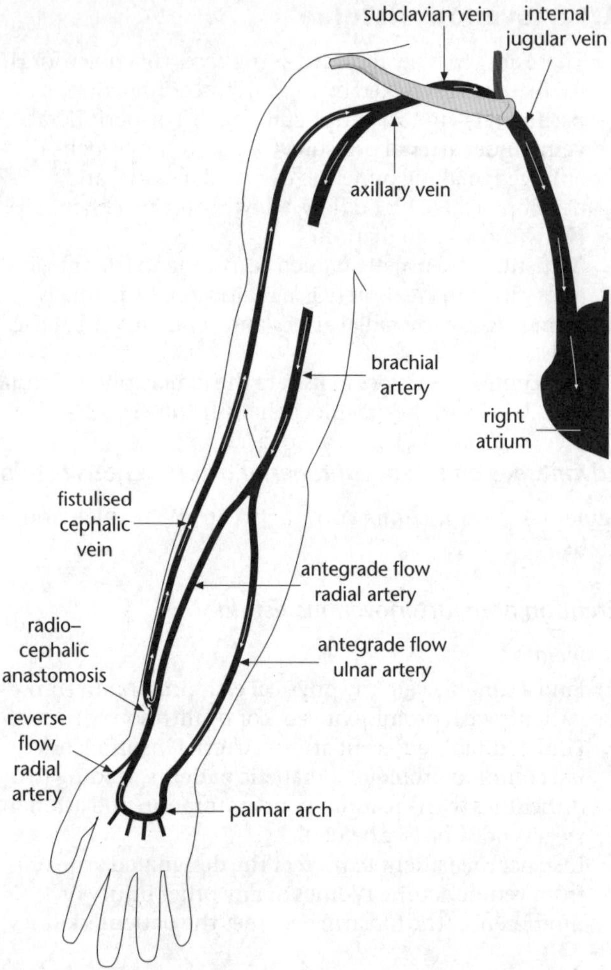

Fig. 3.4C Anatomy of a right radiocephalic fistula. The fistula is formed using the radial artery and the cephalic vein. With time, both the artery and vein will enlarge. Diversion of blood into the cephalic vein may mean that the blood supply to the hand comes entirely by way of the ulnar artery with retrograde flow distal to the radiocephalic anastomosis.

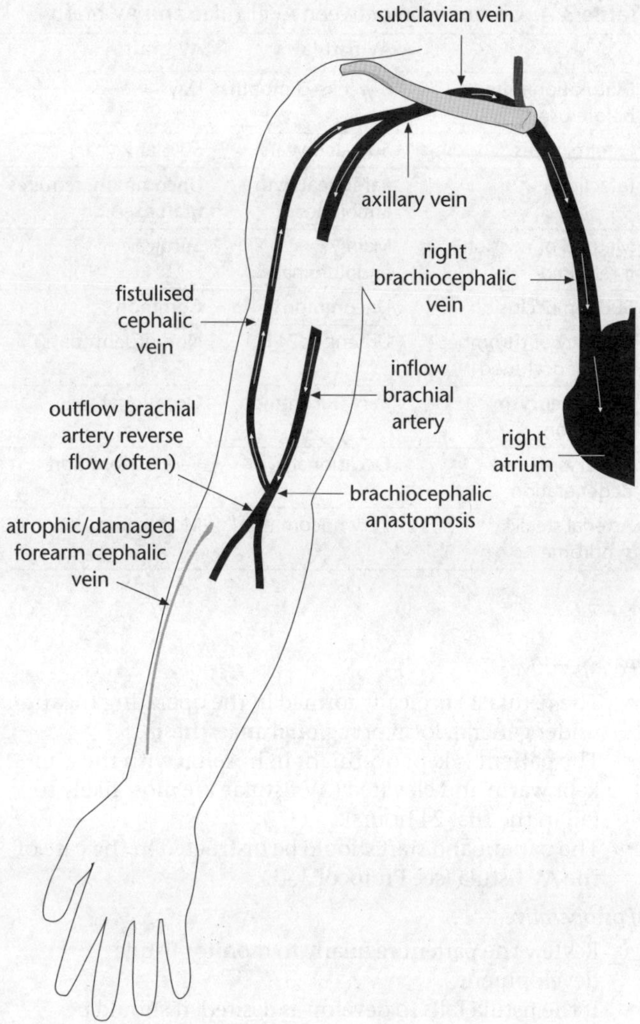

Fig. 3.4D Anatomy of a brachiocephalic fistula. In the example shown, a radiocephalic fistula was not possible due to damage to the distal cephalic vein. The extent of flow through the fistula can cause retrograde flow in the distal brachial artery during diastole.

Table 3.4 Comparison between AV fistulae and AV grafts

	AV fistula	**AV graft**
Maturation time before use	6 weeks–3 months	Days–weeks
Patency rates	50% at 5 years	50% at 2 years
Infection	Rare; treat with antibiotics	Uncommon; requires graft excision
Method of revision if required	Mainly endoluminal	Surgical
Sudden occlusion	Uncommon	Common
Urgency of thrombectomy if occluded	Urgent (<24 h)	Non-urgent (days)
False aneurysm formation	Very uncommon	Occasional
Aneurysmal degeneration	Occasional	Very uncommon
Arterial steal syndrome	Very uncommon	More common

Perioperative

- The fistula is surgically formed in the operating theatre under general, local or regional anaesthesia.
- The patient is kept overnight in hospital with the arm kept warm and elevated (AV fistulae are most likely to fail in the first 24 hours).
- The patient and staff should be instructed in the care of the AV fistula (see Protocol 3.4F).

Postoperative

- Review the patient regularly to monitor fistula development.
- If the fistula fails to develop as desired, it should be assessed with duplex ultrasound by an experienced operator. Contrast fistulography is occasionally needed to diagnose a problem.

- Any significant fistula problem that is identified should then be corrected by endoluminal or open techniques.
- When the fistula is deemed mature (i.e. large enough to be readily accessed and old enough to allow the vein wall to 'arterialise'—usually >4 weeks), it should initially be accessed by an experienced dialysis nurse. Training for self-cannulation should only be attempted after the fistula is deemed to be mature.

PROTOCOL 3.4F

How to care for an arteriovenous fistula

Instructions for staff

- No IV lines, venipuncture, BP cuffs etc. on the arm with the fistula.
- The fistula must only be accessed for the purpose of HD.
- Avoid dehydration/low BP.

Instructions for the patient

- Avoid tight clothing or constriction of the arm/shoulder with the fistula.
- Do not sleep on arm with the fistula.
- Monitor fistula for the presence of a thrill 3 times per day—contact the local HD centre or dialysis ward promptly if the thrill is not detectable.

Complications of AV fistulae

Early complications (within the first 48 hours)

- Occlusion: due to a surgical technical problem; hypotension/dehydration; or inadequate vessels.
- Bleeding, haematoma.

Intermediate-term complications (days–weeks after creation)

- Occlusion: due to hypotension/dehydration or stricture.
- Failure to develop (in order of importance): cephalic vein stricture; anastomotic stricture; inadequate inflow artery.

- Nerve injury (especially in the distribution of the distal radial nerve).
- Arterial steal syndrome, which:
  - is hand ischaemia caused by excessive fistula flow
  - is more common in diabetic patients and those with brachiocephalic fistulae
  - presents, in mild form, as forearm 'claudication', and in severe form, as ischaemic rest pain in hand, especially during dialysis
  - may settle with time or get worse

Long-term complications (weeks–months after creation)

- Occlusion: due to hypotension/dehydration or stricture.
- Poor flow (may be indicated by excessive suction on the arterial limb of the dialysis circuit or excessive recirculation): usually an 'inflow' problem (arterial, anastomotic or distal venous stenosis).
- High venous return pressures: usually an 'outflow' problem (mid or distal cephalic vein stenosis, central vein stenosis).
- Arterial steal syndrome (as above).
- Generalised aneurysmal degeneration, especially in brachiocephalic fistulae.
- Localised aneurysm formation: may be more common with repeated puncture of the same site.
- High output cardiac failure (particularly with brachiocephalic fistulae).

Postoperative bleeding

- This must *not* be managed with pressure dressings or strong compression, which may occlude the fistula.
- Correct treatment includes elevation of the limb as high as possible; gentle, intermittent manual pressure over the bleeding point; early return to theatre for evacuation of haematoma and haemostasis.

Failure of fistula development

See Protocol 3.4G for an assessment of the dysfunctional AV fistula.

PROTOCOL 3.4G

How to assess a dysfunctional arteriovenous fistula

History from patient and staff

- Is it a problem of high 'suction' pressures on the arterial limb of the HD circuit (usually an 'inflow' problem) or of high return pressures in the venous limb (usually indicating an 'outflow' venous stenosis).

Clinical assessment

- Does the fistula collapse with elevation (a high-flow fistula with an outflow problem may not collapse)?

Fistula imaging

- Duplex ultrasound will accurately identify the problem in >80% of cases (with an experienced operator).
- Diagnostic fistulography is occasionally needed.

Intervention

- Most cases (80%) of failed fistula development can be treated endoluminally; the rest will need open revision or hybrid procedures.
- About 5% of fistulae that fail to develop cannot be fixed and a new fistula must be fashioned or alternative ESRF treatment used.

Acute thrombosis of an AV fistula

- For an AV fistula, treatment must take place within 24 hours before thrombophlebitis destroys the fistula vein.
- Intervention involves thrombectomy plus correcting any underlying problem, such as outflow stenosis.

Arterial steal syndrome

- Management depends on severity and timing.

- Early symptoms suggest a significant problem as the steal usually gets worse with fistula development.
- Mild hand claudication is usually benign but rest pain could herald digital ischaemia.
- If steal is severe, the fistula can be 'choked' with a dacron cuff. Usually, however, a problem steal will require the fistula to be ligated.

Generalised aneurysmal degeneration

- This is due to excessive fistula development, which may result in a 'giant fistula' with a feeding artery up to 2 cm in diameter.
- It can produce high-output cardiac failure.
- It is principally seen with brachiocephalic fistulae.
- There is the danger of catastrophic bleeding.
- It is best treated with ligation of the fistula.

Localised aneurysmal dilatation

- This is often due to repeated puncture of same segment of vein.
- There is the risk of thrombosis within aneurysmal segments.
- It is prevented by rotation of the puncture sites.
- Aneurysms can be surgically plicated.

Recirculation within an AV fistula

Chapter 3.7 discusses recirculation.

Arteriovenous grafts

- These are a synthetic graft (e.g. PTFE) connecting an artery and vein.
- They are most commonly placed in the forearm between the brachial artery and basilic vein, and less commonly in the arm or in the thigh. Rarely, they can be placed on the chest wall using the internal jugular vein for outflow.

- They are a poor substitute for an AV fistula (see Table 3.4) but have a significantly reduced maturation time (can sometimes be used within 24 hours after formation).

Complications of AV grafts

Poor function

- Complications are more commonly an 'outflow' problem at the venous end causing high return pressures in the venous limb of the circuit.
- They require surgical revision, interposition graft or patch repair.
- Assessment and management is similar to that of AV fistulae, although duplex ultrasound is less useful for assessment (due to poor transmission of the ultrasound waves through the prosthetic material). Contrast fistulography is, therefore, more commonly recommended for troubleshooting.

Occlusion

- Delayed management is possible (unlike with AV fistulae).
- This requires thrombectomy of the graft ± revision of stenosis.

Infection

- AV graft infections should be treated aggressively with IV antibiotics (see Chapter 3.7).
- The risk of necrotic perforation is high in cases of staphylococcal AV graft infection and consideration should be given to early revision of the graft if sepsis remains a problem for more than 24–48 hours.

Chapter 3.5
Anticoagulation for haemodialysis

HIGHLIGHTS

- Contact with plastic tubing, the dialysis membrane and air in the HD circuit stimulates the clotting cascade.
- Excessive clotting in the dialysis circuit and filter may mean the tubing and filter need to be discarded; in adults, this can mean the loss of 120–250 mL of blood.
- Clotting within an HD circuit can be minimised through appropriate use of anticoagulation therapy.
- If the patient is not receiving oral anticoagulant therapy, periodic anticoagulation is normally given during the dialysis treatment.
- The prescription for anticoagulation therapy must be decided in consultation with the medical team and be documented on the medication chart.

Contraindications to the use of anticoagulation therapy during dialysis

Absolute and relative contraindications include:

- possibility of active bleeding (e.g. acute neurological event, acute pancreatitis)
- intercurrent therapeutic anticoagulation
- the presence of coagulopathy
- abnormal liver function tests
- platelet count $<150 \times 10^9$/L
- antiplatelet therapy

- preoperation
- <48 hours after an invasive procedure, including angiograms and renal biopsy

In the above circumstances, heparin-free dialysis may be necessary. This involves flushing the dialyser with 100 mL of normal saline 0.9% every 30–60 minutes. Blood flows need to be kept as high as possible to reduce the risk of clotting.

Anticoagulation protocols for haemodialysis

Anticoagulants

Three different anticoagulants are routinely used for periodic anticoagulation during dialysis (see Protocol 3.5):

- unfractionated heparin
- low-MW heparin (LMWH)
- synthetic heparinoids

Low-molecular-weight heparins

LMWH are regarded as more effective for dialysis with a lower risk of HIT than with unfractionated heparin. They are more expensive and should be considered in patients who have ongoing evidence of clotting despite unfractionated heparin at a maximal dose (2000 U/h).

Synthetic heparinoids

Synthetic heparinoids are also expensive and are usually reserved for patients who have developed immune-mediated HIT with either unfractionated heparin or LMWH (see Chapter 2.3).

PROTOCOL 3.5

How to prescribe anticoagulation during haemodialysis

Unfractionated heparin

- Initial HD: loading dose of 250–500 U followed by an infusion rate of 250–500 U/h.
- The rate of infusion is increased at each dialysis by 200–500 U/h increments until adequate anticoagulation is reached (2000 U/h is maximum level).

Low-molecular-weight heparins

Dalteparin (Fragmin)

- In patients with no bleeding risk and HD duration <4 h:
  - bolus dose prior to dialysis initiation of 2500–5000 U *or* a loading dose of 5–10 U/kg and an infusion rate of 4–5 U/kg/h
- In patients with no bleeding risks and HD >4 h:
  - loading dose of 30–40 U/kg and an infusion rate of 10–15 U/kg/h
- In patients with a bleeding risk or acute renal failure:
  - loading dose of 5–10 U/kg and an infusion rate of 4–5 U/kg/h

Enoxaparin sodium (Clexane)

- Use 1 mg/kg in the arterial line at the start of a 4-hour dialysis session.
- If performing convective dialysis therapies (high flux or HDF), inject the Clexane into the venous line as the drug can be cleared from high-flux membranes before anticoagulation is achieved.

Synthetic heparinoids

Danaparoid sodium (Orgaran)

- Initial dialysis: 3750 U IV bolus dose prior to dialysis; if patient is less than 55 kg, then give 2500 U.
- Second dialysis: same dose as at initial dialysis.
- Third dialysis: 3000 U IV bolus prior to dialysis, providing there are no fibrin threads in the venous chamber; if the patient is less then 55 kg, then give 2000 U.

- When using synthetic heparinoids, the anti-Xa levels need to be checked prior to the second dialysis and each subsequent dialysis to determine the appropriate dosing levels. During dialysis the plasma anti-Xa level should be 0.5–0.8 U/mL. The routine for testing is predialysis and during dialysis at 30 minutes and 4 hours.

Anticoagulation regimens

For all anticoagulation regimens, the doses need to be individualised for each patient. The effectiveness of anticoagulation can be assessed by:

- visual inspection for clotting in the dialyser and bubble traps—clotting will require an increase in the dose of anticoagulant
- assessing haemostasis at the needle puncture sites at the end of dialysis—this is normally obtained within 10 minutes by firm pressure; excessive bleeding times may require either a reduction in the anticoagulation dose or earlier cessation of the infusion (if unfractionated heparin is used)

Chronic anticoagulation therapy for access patency

- Some patients with recurrent thrombosis of an AV fistula or graft may require oral antiplatelet or anticoagulation therapy to reduce the risk of access thrombosis.
- Antiplatelet treatment is usually given as low-dose aspirin (100–150 mg per day). These patients may be given conventional levels of anticoagulation therapy during HD using one of the above protocols.
- Low-dose oral anticoagulation therapy with warfarin is used occasionally (usually aiming for an INR of 1.5–2).

Chapter 3.6
The first haemodialysis treatment

HIGHLIGHTS

- Patients undergoing their first HD treatment are at significant risk of dialysis disequilibrium if the metabolic abnormalities of uraemia are corrected too rapidly.
- Dialysis disequilibrium:
  - is thought to be caused by cerebral oedema related to the rapid removal of uraemic solutes from the ECF compartment
  - manifests as headaches, confusion and blurred vision; in severe cases, it can be associated with seizures
- Thus, a graded approach is typically used (i.e. several short treatments of increasing intensity over consecutive days).
- See Protocol 3.6 on how to manage the first HD.

PROTOCOL 3.6

How to manage the first haemodialysis

Determine whether the patient poses a potential cross-infection risk.

- Check that viral serology (e.g. for hepatitis B, C and HIV) has been sent or checked. Some units also require screening for other sources of cross-infection risk (e.g. skin colonisation with resistant organisms—MRSA, VRE).

- If the results of viral serology are not known, then the dialysis machine will require stringent disinfection afterwards (e.g. with Dialox).
- If a temporary CVC is being used, ensure that the tip position has been confirmed radiographically (not necessary for femoral catheters).

Determine the dry weight of the patient (see also Protocol 3.3E) and whether any variances to standard anticoagulation exist (see Chapter 3.5).

The initial HD regimen for adults is as follows:

- dialyser—1.6 m^2, low flux, synthetic membrane and steam sterilised
- blood flow—150 mL/min, increasing by 25–50 mL/min at each subsequent dialysis
- time—2 hours, increasing by 1 hour at each subsequent dialysis until 4 hours is reached
- dialysate flow—500 mL/min
- dialysate temperature—36°C
- dialysate sodium:
  - serum sodium levels >125 mmol/L—dialyse against a sodium concentration of 140 mmol/L
  - serum sodium levels <125 mmol/L—dialyse against a sodium concentration no higher than 15 mmol/L above the plasma level
  - serum sodium levels >155 mmol/L—dialyse against a sodium concentration no lower than 15 mmol/L below the plasma level

Chapter 3.7
Complications during haemodialysis

HIGHLIGHTS

The complications of HD may be categorised as follows:

- hypotension
- sepsis
- hypoxaemia
- membrane reactions
- access recirculation
- air embolus
- haemolysis
- dialysis disequilibrium syndrome (see Chapter 3.6)

Intradialytic hypotension

Hypotension is the most common problem encountered during a dialysis session. Protocol 3.7A shows how to manage hypotension during HD.

Intradialytic sepsis

- Sepsis of any cause is common in patients on dialysis (see Chapter 2.6).
- Sepsis in HD patients is most commonly related to the dialysis access (particularly if an HD catheter is in use; see Chapter 3.4).
- Protocol 3.7B outlines the management of the patient with sepsis during HD and Protocol 3.4C the assessment and management of HD catheter-related sepsis.

Hypoxaemia during dialysis

- This is a well-described phenomenon that can occur in the absence of overt pulmonary causes, such as pulmonary oedema, pneumonia or pulmonary embolism.
- It is particularly hazardous in the patient with borderline pulmonary function.

Possible causes

- Hypoventilation due to the relative alkalaemia from the use of a bicarbonate-based dialysis.
- Pulmonary leucostasis (particularly with cellulosic membranes).

Treatment

- Supplemental oxygen.
- Avoid excessive dialysate bicarbonate.

Membrane reactions

- These arise from the interaction between the blood constituents and the membrane.
- There are two types of membrane reactions—type A and type B. Table 3.7 gives a comparison of these reactions.

Access recirculation

- Access recirculation occurs when dialysed blood returning to the patient from the dialyser is drawn back into the dialyser circuit.
- The effectiveness of HD is subsequently reduced.

Assessment

See Protocol 3.7C for an assessment and management of access recirculation.

PROTOCOL 3.7A

How to manage hypotension during haemodialysis

Immediate management

- Slow or stop UF and blood flow.
- Place the patient in the Trendelenberg position.
- Give a 200–300 mL bolus dose of IV saline.

Evaluate for the underlying cause

- Is hypotension a manifestation of shock due to:
    - myocardial infarction, arrhythmia or pericardial tamponade?
    - pulmonary embolism?
    - sepsis (see Protocol 3.7B)?
    - haemorrhage?
- Is the rate or extent of fluid removal during dialysis excessive?
    - Is there excessive interdialytic weight gain (i.e. more than 3 kg between dialysis treatments)?
    - Is the dry weight set correctly (see Protocol 3.3E)?
    - Is the UF rate incorrect or erratic (this problem has been largely overcome with the use of UF controllers)?
    - Is the dialysate sodium set too low (see Chapter 3.3 and Table 3.3B)?
- Is there impaired haemodynamic compensation?
    - Is there an underlying cardiac disease (impaired ability to compensate for intravascular fluid shifts)?
    - Is the patient taking antihypertensive medications or beta-blockers (often withheld on the day of dialysis)?
    - Has the patient had a meal prior to or during dialysis (this promotes splanchnic blood flow and may precipitate hypotension in patients with cardiovascular disease)?
    - Does the patient have autonomic neuropathy?

Prevention of hypotension during dialysis

- Reduce interdialytic weight gain (salt and fluid restrict the patient).
- Reset the dry weight (see Protocol 3.3E).
- Treat anaemia (see Chapter 2.2).

- Withhold antihypertensive medications on dialysis days.
- Avoid meals during or just prior to dialysis.
- Consider adjustment of the dialysate sodium or reduction of the dialysate temperature (see Table 3.3B).
- Use a UF controller.
- Consider using a period of isolated UF (this causes less haemodynamic instability).
- Consider using caffeine. Adenosine release by subclinical visceral organ ischaemia during episodes of hypotension may lead to vasodilatation and further hypotension, resulting in a vicious cycle. Caffeine, an adenosine agonist, may ameliorate this effect.
- Consider the use of midodrine, an alpha-adrenergic agonist, for patients in whom hypotension is refractory to the above measures.

PROTOCOL 3.7B

How to manage the patient with sepsis during haemodialysis

- Suspect sepsis if the patient on dialysis develops fever, chills or hypotension while on HD.
- Resuscitate first with oxygen therapy; treat hypotension if present.
- Look for the cause:
  - If the patient has a dialysis catheter, is there evidence of an exit site infection (i.e. purulence or cellulitis at the point of skin insertion)?
  - Is there cellulitis around the AV fistula or graft?
  - Assess for non-access related sources of infection in the absence of the above.
  - Take blood samples for aerobic and anaerobic culture.
- Treatment:
  - Empirical broad-spectrum parenteral antibiotics in cases of suspected access-related sepsis (e.g. gentamicin and either dicloxacillin or cephazolin). Vancomycin should be used if there is a history of MRSA colonisation.
  - All non-tunnelled lines should be removed in all cases of suspected access-related sepsis and the catheter tip sent for bacterial culture. Patients with a tunnelled dialysis catheter who are febrile may be given a 24–48 hour trial of antibiotics, failing which the catheter should be removed. Signs of tunnel infection, persistent sepsis or fungal septicaemia warrant catheter removal.
- If the patient is found to be bacteraemic, a minimum of 7 days of antibiotic therapy is indicated. A 14-day duration of therapy is recommended for patients with *Staphylococcus aureus* bacteraemia.
- AV graft infections should be treated aggressively with IV antibiotics. In staphylococcal AV graft infection, the risk of necrotic perforation is high; therefore, consideration should be given to early revision of the graft if sepsis remains a problem for more than 24–48 hours.

Prevention of access-related sepsis

Dialysis catheters

- Minimise the use of non-tunnelled dialysis catheters to less than 3 weeks wherever feasible (CARI, K/DOQI guidelines).
- Treat exit site infections promptly—see Protocol 3.4B.
- Keep the catheter insertion site clean and dry.
- Apply mupirocin or povidine-iodine ointment to the exit site.
- Keep the vascular access insertion site clean and dry.
- Consider using an antibiotic-impregnated catheter.

General measures

- Observe strict infection control procedures within the dialysis unit (i.e. hand-washing, universal precautions).
- Consider antimicrobial prophylaxis (i.e. prophylactic antibiotic therapy at the time of access creation/insertion and for invasive procedures while the vascular access device is present).
- Consider elimination of nasal carriage of MRSA with mupirocin ointment (of unproven benefit).

PROTOCOL 3.7C

How to assess and manage access recirculation

Measurement

Measure recirculation within the first 30 minutes of dialysis by either the conventional method or the dilutional method.

Conventional method (three needle method)

- The urea level is measured in blood samples collected from the:
  - peripheral circulation (*P*)
  - arterial limb of the HD circuit (*A*)
  - venous limb of the HD circuit (*V*)
- Access recirculation (*AR*) can then be calculated as:

$$AR\% = \frac{(P - A)}{(P - V)} \times 100\%$$

- An abnormal value for this method is generally regarded as greater than 15% recirculation.

Dilutional methods

- These are faster and more accurate but require specialised equipment.
- They are based upon either thermodilution or ultrasound dilution of saline injected into the venous line.
- Recirculation values of more than 5% using these methods are considered abnormal.

Management

- Correct needle placement technique—the tips of the arterial and venous needles should ideally be separated by at least 4 cm.
- Are there signs of poor fistula flow (e.g. poor flows attainable during dialysis, high suction pressures in the arterial line)?
  - Assess access 'inflow' using either duplex ultrasound (for AV fistula) or radiocontrast (AV graft). See Protocol 3.4G.
- Are there signs of fistula outflow obstruction (suggested by high pressure in the venous line)?

  - Assess access 'outflow' using either duplex ultrasound (for AV fistula) or radiocontrast (AV graft). See Protocol 3.4G.

Treatment

- Temporarily increase the duration of dialysis.
- Revise or replace HD access.

Table 3.7 Comparison of membrane reactions

	Type A	Type B
Frequency	Uncommon	Common
Severity	Major	Minor
Timing	Immediate (within minutes of commencing dialysis)	Delayed (15–30 minutes after starting dialysis)
Clinical course	Worsens with continued dialysis	Resolves with continued dialysis
Symptoms	Anaphylactoid (dyspnoea, burning sensation, angioedema, urticaria, cough, diarrhoea, headache, rhinorrhoea, lacrimation, sense of impending doom, hypotension, death)	Back pain, chest pain, dyspnoea, nausea and vomiting. Anaphylaxis extremely rare
Associations	Reusing dialysers sterilised with ethylene oxide, use of acrylonitrile-sodium methallylsulfonate (AN-69) dialysers with ACEIs, heparin, acetate	
Pathogenesis	Bradykinin generation, allergic diathesis	Complement generation
Treatment	Stop dialysis Do not return blood Resuscitation with fluids, bronchodilators, adrenaline	Continue dialysis No specific therapy
Prevention	Rinse dialyser well Consider dialyser reuse Avoid ethylene-glycol sterilised dialysers Avoid ACEIs	Consider dialyser reuse (dialysis membrane coated with patient's plasma proteins may prevent reactions during next dialysis)

Air embolus

- This is an uncommon complication of HD nowadays due to the use of air detectors in HD apparatus.
- See Protocol 3.7D for an assessment and management of suspected air embolism during HD.

Haemolysis

Massive haemolysis may occur in patients on dialysis and is a potentially life-threatening complication.

PROTOCOL 3.7D

How to assess and manage suspected air embolism during haemodialysis

Symptoms

Symptoms depend on the position of the patient.

- If sitting up, air migrates upwards into the cerebral venous system, causing unconsciousness, coma and death.
- If recumbent, air migrates into the right heart and propagates into the lungs, causing dyspnoea, and may enter the left side of the heart, embolising and causing acute neurological and cardiac dysfunction. It is rapidly fatal.

Signs

- Foam in the venous part of the dialysis circuit.

Treatment

- Clamp the venous line immediately.
- Lie the patient in the left lateral position with the head down.
- Call a cardiac arrest and resuscitate the patient with 100% oxygen.
- Percutaneous aspiration of air from the heart may be required.

Prevention

- Modern equipment has air detectors that largely prevent air embolism.
- Proper attention to line connections is necessary.

Assessment and management

See Protocol 3.7E for an assessment and management of suspected haemolysis during HD.

PROTOCOL 3.7E

How to assess and manage suspected haemolysis during haemodialysis

Symptoms

- Acute dyspnoea, chest pain and back pain.

Signs

- Port-wine discolouration of blood in venous line.
- Pink discoloration of the plasma in centrifuged blood and/or falling haemoglobin level.

Treatment

- Immediate clamping of the dialysis lines and cessation of dialysis.
- Urgent measurement and management of hyperkalaemia (due to release of potassium from haemolysed red blood cells).

Investigations

Investigate for possible causes:

- contamination of dialysis water by chloramine, copper, zinc, nitrates, nitrites
- overheated or hypoosmolar water
- kinked, blocked or otherwise defective tubing
- formaldehyde or bleach used in dialyser reprocessing
- other causes, as in a non-dialysis patient (e.g. drug-induced, autoimmune, hypersplenism, haemolytic uraemic syndrome)

Dialysis disequilibrium

See Chapter 3.6.

Chapter 3.8
Haemodialysis in children

HIGHLIGHTS

- HD in children generally provides better clearances than does PD.
- HD is frequently the preferred mode of chronic dialysis delivery in the young adult.

Unique aspects of haemodialysis in children

Staff

- It is important to have dedicated experienced nursing staff for the delivery of HD to children. Recognition of the volume status or stress in the child on dialysis is difficult and requires experience and the ability to respond rapidly to haemodynamic changes.

Immunisations

- As most children on dialysis are awaiting a transplant, this is an important time to screen for immunisation status and ensure the child is fully immunised prior to transplant.
- In addition to routine immunisations, children should be immunised against hepatitis B and *Pneumococcus*.

Development

- Children with renal failure have multiple issues that may affect their development, including time away from school, problems related to uraemia and specific defects related to the underlying aetiology of their renal disease.

- It is important to have full developmental assessments of children with renal failure, particularly in young children. A multidisciplinary team, including a psychologist, educator, dietitian, nurse, physician and occupational therapist, provides this best.

Dose of dialysis

- Because most children have a small BSA, the delivered dialysis dose is frequently high and the *Kt*/*V* targets for adults are easily exceeded.
- Regular monitoring of the dialysis dose is necessary. A decline in the delivered dialysis dose may reflect either a need for more dialysis in a growing child or failure to deliver dialysis because of recirculation or obstruction in the fistula.

Dialysis equipment

- Newer dialysis machines provide close volumetric monitoring of dialysis. This is crucial in children and these machines should be used for children in preference to older machines.
- Dialysis lines should have the minimal volume possible to reduce extracorporeal volume.
- Dialysis filters are chosen to provide maximum clearance for the child's size.

Special circumstances

- Small children who weigh less than 15 kg frequently require blood/albumin priming of the circuit to minimise the haemodynamic effects of dialysis.
- Children with a high osmotic load at the commencement of dialysis (urea level >50 mmol/L) and children with associated liver failure are at risk of cerebral oedema from disequilibrium. If continuous HD is not available, a mannitol infusion in parallel with dialysis is recommended.

Haemodialysis in children with acute renal failure

- Intermittent HD is the preferred form of acute dialysis where rapid clearance is required, such as in acute toxic poisoning, or for the correction of hyperkalaemia.
- In most cases where continuous HD is available, the gradual correction with this method initially is preferable. Transfer to conventional HD is often performed when the child is stable.
- Conventional HD may be preferable in the case of the removal of toxins, such as ethylene glycol, where rapid clearance is of therapeutic importance.
- Similar to adults, acute vascular access in children is established using a dual lumen dialysis catheter placed into a major vein or right atrium.

Initial dialysis

- Because of the risk of disequilibrium it is often best to start with a limited dialysis run, without fluid removal unless necessary.
- Daily dialysis can then be instituted, gradually increasing the duration with each dialysis.
- Once the child is stable, he/she can then be moved onto a 3–4 times a week dialysis schedule.
- It is important to establish the dry weight and aim to remove fluid to achieve this weight. Correction of hypoalbuminaemia and ensuring adequate intravascular volume prior to dialysis are important for adequate fluid removal. Anticoagulation is often required and heparin is used to prevent thrombosis.

Stopping dialysis

- In children where a reversible cause is suspected, the predialysis serum creatinine level can be monitored to assess renal recovery.

- Where recovery is expected, it is reasonable to maintain the child above their dry weight to prevent prerenal ATN prolonging the recovery.

Chronic haemodialysis in children

Long-term haemodialysis access in children

In children likely to need long-term HD, early assessment by a vascular surgeon and fistula placement are important. Fistulae have significantly less morbidity and allow a better quality of life.

Complications of long-term haemodialysis

- Children on long-term HD may develop autonomic neuropathy and find it difficult to deal with fluid overload or dehydration.
- Maintaining suitable vascular access becomes an increasing problem with time. In this case, moving to daily dialysis or high-flux dialysers may improve the dialysis dose achieved with the existing access.

Part 4
Chronic peritoneal dialysis

The purpose of this part is to provide a summary of the general principles of chronic PD.

Chapter 4.1
General principles

HIGHLIGHTS

- PD utilises the peritoneal membrane as a natural 'dialysis filter'.
- PD solution is instilled into the peritoneal cavity and acts as the 'dialysate'.
- PD has lifestyle and medical advantages over HD.
- The outcome of patients on chronic PD is equivalent to that of those on HD during the first 3 years but long-term technique survival after 5 years is low.

Basic principles

The basic system of chronic PD is shown in Figure 4.1.

- PD utilises the peritoneal membrane as a dialysis filter.

- A PD catheter is inserted surgically into the peritoneal cavity.
- PD solution is infused into the peritoneal cavity via the catheter (called an *exchange*) and remains there for up to 6–10 hours (called a *cycle*), during which time uraemic toxins and fluid are removed by diffusion and UF across the peritoneal membrane.
- A typical patient performs 4–5 exchanges per day, 7 days a week, and this produces sufficient solute and water removal to remain free of uraemic symptoms, and to provide a GFR equivalent to about 10 mL/min.

Advantages of PD compared to HD

- Home-based—can be performed in remote geographical locations.
- Self-care dialysis—no helper required unless patient has physical or social problems.
- Simple to learn and perform—training period 1–2 weeks.
- Greater patient independence and travel—holidays easier to arrange.
- Flexible schedule—not restricted to hospital rules and appointments.
- As dialysis is continuous the patient is allowed a more liberal dietary and fluid intake.
- There are no haemodynamic complications or need for chronic vascular access.
- It is especially suited to patients with cardiovascular disease or the elderly.
- PD preserves residual renal function better than does HD.
- Patient survival is better or equivalent to that with HD during the first 3 years.
- Patients on PD who subsequently receive their first cadaveric kidney transplant have a slightly lower risk of delayed graft function but long-term graft survival is unchanged.

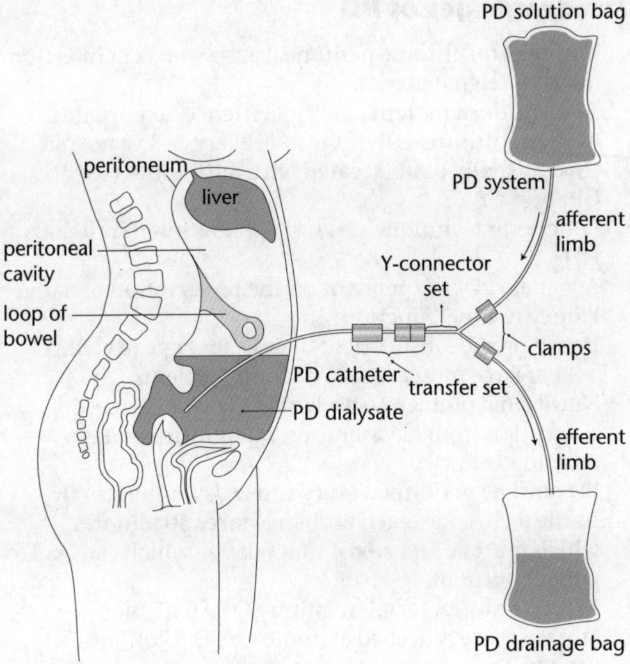

Fig. 4.1 Basic system of chronic peritoneal dialysis. The components of the system include the PD catheter, the transfer set and the Y-set double bag system (consisting of the Y-connector and twin PD bags—one drainage bag and one fresh PD solution bag, which come prepackaged)

Disadvantages of PD

- The need for chronic peritoneal access—risk of infection due to foreign material.
- Two-thirds of patients may experience an episode of peritonitis (usually 1 episode every 2–3 years) but this is usually easily treated with antibiotics (see Chapter 4.8).
- Long-term technique survival remains inferior to that of HD.
- Adequate PD is dependent on the preservation of native kidney residual function.
- It may provide insufficient dialysis for large (e.g. BMI >35 or large muscle mass) or anuric patients.
- Nutritional problems—weight gain due to glucose absorption from PD solutions; malnutrition due to protein losses.
- PD must be performed every single day using strictly sterile technique (each exchange takes 30 minutes, which must be repeated 4 times daily), which can lead to patient burnout.
- PD can result in social isolation of the patient.
- Storage space is needed at home for PD solutions—usually 2 m^3.

Outcome of patients on chronic PD

- One-quarter of all chronic dialysis patients perform PD.
- Patient survival and death-censored technique survival are both about 50% at 3 years.
- Only a small percentage of patients (5%) remain continuously on PD for 5 years or more.
- Social factors and infectious complications are the main reasons for technique failure.

Choice of PD as a dialysis modality

- Because of the advantages, PD should be considered as the initial chronic dialysis modality treatment in all patients unless there are contraindications (see Table 1.5).

- Determining whether a predialysis patient with CKD is suitable for PD involves multidisciplinary assessment by nursing, social worker, medical and surgical personnel specialised in the management of chronic PD patients (see Protocol 4.1).

PROTOCOL 4.1

How to assess patient suitability for chronic peritoneal dialysis

- Assess whether there are factors that would favour PD over home HD:
  - medical—difficult vascular access, severe cardiovascular disease
  - personal—remote geographic location, living alone
- Determine whether there are *absolute* or *relative* contraindications to PD:
  - multiple previous abdominal operations, colostomy, severe respiratory disease
  - large body size or anuric (patient *may* not achieve adequate dialysis with PD)
- Determine whether there are any potential problems that need evaluation before starting PD:
  - Will the patient need assistance because of physical or social problems?
  - If the patient has a hernia, this can be corrected at the time of PD insertion.
  - If the patient has poor eyesight, refer to an ophthalmologist.

References

1. Johnson D. Chapter 6. Peritoneal dialysis. ANZDATA Registry 2003 Report, Australia and New Zealand Dialysis and Transplant Registry, Adelaide, SA, 2004.
2. Gokal R. Peritoneal dialysis in the 21st century: an analysis of current problems and future developments. J Am Soc Nephrol 2002; 13: S104–S116.
3. McDonald SP, Collins JF, Johnson DW. Obesity is associated with worse peritoneal dialysis outcomes in the Australian and New Zealand patient populations. J Am Soc Nephrol 2003; 14: 2894–901.

Chapter 4.2
Peritoneal membrane physiology

HIGHLIGHTS

- The peritoneal membrane has a large surface area.
- An osmotic and concentration gradient is created by the presence of the PD solution in the peritoneal cavity, allowing the transfer of solutes and water from the peritoneal capillaries into the PD solution.
- The rate of solute and water transport can be estimated periodically in patients using the peritoneal equilibration test (PET).

Normal anatomy

- The peritoneum is a thin, semipermeable, serosal membrane that lines the peritoneal cavity. It has a large surface area (0.55 m^2) and is divided into two parts:
  - the parietal peritoneum, which lines the walls of the abdominal cavity (20% of the total peritoneal surface area)
  - the visceral peritoneum (may be less important for PD), which lines the gut and visceral organs (80% of the total peritoneal surface area)
- In normal states, the peritoneal cavity contains 100 mL or less of fluid. A normal adult can tolerate 2 L or more of fluid in the cavity without pain or alteration to respiratory function.
- In males, the peritoneal cavity is closed but in females it is continuous with the Fallopian tubes. Peritoneal fluid leakage is prevented in females as the Fallopian tube

openings are collapsed, although PD fluid can, rarely, become blood-stained during a menstrual period.

- The total peritoneal blood supply is estimated to be 60–100 mL/min. Under basal conditions, only about 25% of the peritoneal capillaries are perfused.

The peritoneal membrane in peritoneal dialysis

- During PD, solutes and water must move across five layers from the blood to the peritoneal cavity, which contains the PD solution. These are the: capillary endothelium; endothelial cell basement membrane; interstitium; basement membrane; and mesothelium (see Fig. 4.2).

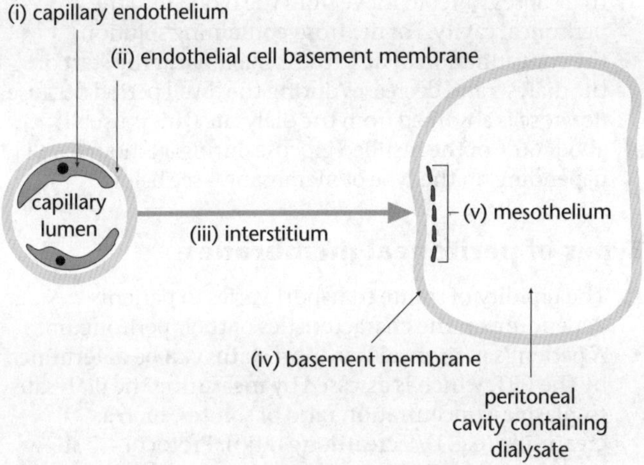

Fig. 4.2 The layers of the peritoneal dialysis membrane. Solute and water must cross five barriers to move from the blood into peritoneal cavity: (i) capillary endothelium; (ii) endothelial cell basement membrane; (iii) interstitium; (iv) basement membrane of the mesothelium; and (v) mesothelium of the visceral peritoneum

- The number of perfused peritoneal capillaries, rather than the total anatomical surface area, is the most important determinant of peritoneal solute and water transport.

Mechanisms to move water and solutes across the peritoneal membrane

- Solutes and water move across the peritoneum by diffusion/convection and UF (see Tables 4.2A and 4.2B).
- PD solutions do not contain small solutes (e.g. urea, creatinine, potassium), which creates a concentration gradient across the membrane, thus allowing for solute removal.
- PD solutions contain an osmotic agent (usually dextrose or, less commonly, colloid or amino acids), which creates an osmotic pressure gradient across the membrane that causes water to move from the blood into the peritoneal cavity. For dextrose-containing solutions, the concentration of dextrose is maximal at the start of the dialysis and decreases during the dwell period because dextrose is absorbed from the dialysate (this is usually about 66% of the instilled glucose during a 6-hour dwell, depending on the type of membrane—see below).

Types of peritoneal membranes

- The rapidity of solute transport varies in patients depending on the characteristics of their peritoneum.
- A patient's peritoneal transport status can be determined by the PET, which is assessed by measuring the dialysate-to-plasma-concentration ratio of solutes, such as creatinine (e.g. D/P creatinine ratio). Protocol 4.2 shows how to perform PET.
- Using the PET, patients are classified into four membrane types according to the rate at which they are able to transport solute (see Table 4.2C).
- The characteristics of a patient's membrane type are not static and vary over time.

Table 4.2A Mechanisms of solute and water transport in peritoneal dialysis

Substance removed	Mechanism
Diffusion Solute (e.g. urea, creatinine, potassium)	• Via mid-size pores. • Most important mechanism of solute transport in PD. • Movement of solutes across the peritoneum from an area of high concentration (blood) to an area of low concentration (peritoneal cavity containing PD solution). • Amount of diffusion is proportional to the concentration gradient. • Solute diffusion is highest at the beginning of the exchange and decreases during the dwell due to loss of the concentration gradient. • Best for clearance of small molecules.
Convection Solute (e.g. protein, sodium)	• Movement of solutes across the peritoneal membrane with water in response to a positive transmembrane pressure. • Less dependent on molecular size.
Ultrafiltration Water	• Movement of water across the peritoneum from an area of *low* osmotic agent concentration (blood) to an area of *high* osmotic agent concentration (the peritoneal cavity containing PD solution) (via the water channel pore aquaporin-1). • The higher the concentration of osmotic agent (in the PD solution), the more water is removed. • UF is highest at the beginning of the exchange when the osmotic gradient is highest, and ceases when dialysate osmolarity has decreased to equal serum osmolarity (due to dextrose absorption). Reabsorption of water occurs if dialysate is allowed to dwell beyond the time past when osmotic equilibrium is reached.

- Membrane type has clinical implications and determines the PD prescription (dwell time for optimal solute clearance and UF). For example, in patients classified with membranes that are high or rapid transporters, solute clearance is excellent but UF is poor because of rapid absorption of glucose (and therefore loss of the osmotic gradient). See Table 4.2C.

Table 4.2B Substances lost and absorbed across the peritoneum

Substances lost into the dialysate	Substances absorbed into the circulation
Small molecules (e.g. urea, creatinine)	Dextrose (if dextrose-containing PD solution)
Some electrolytes (notably potassium)	Calcium
Proteins (up to 5–10 g/d)	
Amino acids, trace minerals, hormones, drugs	

PROTOCOL 4.2

How to perform a peritoneal equilibration test

- Weigh a fresh 2-L 2.5% dextrose PD solution and perform a PD exchange as normal.
- Take a venous blood sample during the dwell to measure serum creatinine and glucose levels.
- Drain out the PD fluid after 4 hours, weigh it and draw a sample of fluid to measure the dialysate creatinine and glucose concentrations.
- The net UF volume is calculated by subtracting the weight of the PD bag before use from the weight after use.
- Divide the dialysate creatinine level by the plasma creatinine level to calculate the D/P ratio.

Table 4.2C Peritoneal membrane types classified by the peritoneal equilibration test

Membrane type	4-h D/P creatinine ratio	Clinical implications
High (or rapid/fast) transporter	0.82–1.03	• Fast solute diffusion • Malnutrition (higher protein loss) • Lower UF • More rapid loss of RRF • Higher patient mortality • Lower technique survival
High average transporter	0.65–0.81	Intermediate
Low average transporter	0.50–0.64	Intermediate
Low (or slow) transporter	0.34–0.49	• Slow solute diffusion • Higher UF • Lower patient mortality • Better technique survival

References

1. Flessner MF. Peritoneal transport physiology: insights from basic research. J Am Soc Nephrol 1991; 2: 122–35.
2. Twardowski ZJ, Nolph KD, Khanna R, et al. Peritoneal equilibration test. Perit Dial Bull 1987; 7: 138–47.

Chapter 4.3
Peritoneal dialysis access and complications

HIGHLIGHTS

- Timely insertion of a PD catheter will allow smooth transition to chronic PD.
- Mechanical complications that occur after catheter insertion may require surgical revision and/or temporary HD support.

Nature of the peritoneal dialysis catheter

The PD catheter is shown in Figure 4.3. It consists of:

- Silastic tubing
- a perforated tip for free flow and drainage of dialysate
- single or double cuffs (a deep cuff in the abdominal wall and a subcutaneous cuff to maintain the catheter position by inducing scar tissue)
- a distal tip which may be straight or curled in shape and is attached externally to a 'transfer set' (usually made of a titanium lock with connecting tube), that is connected to the Y-set (see Fig. 4.1). The transfer set is usually not attached until the patient commences training and is designed to minimise the risk of accidental microbial contamination of the PD catheter. Different types of transfer sets exist and each is designed to suit a Y-set double bag system from specific manufacturers. The transfer is changed in the PD unit every 6 months.

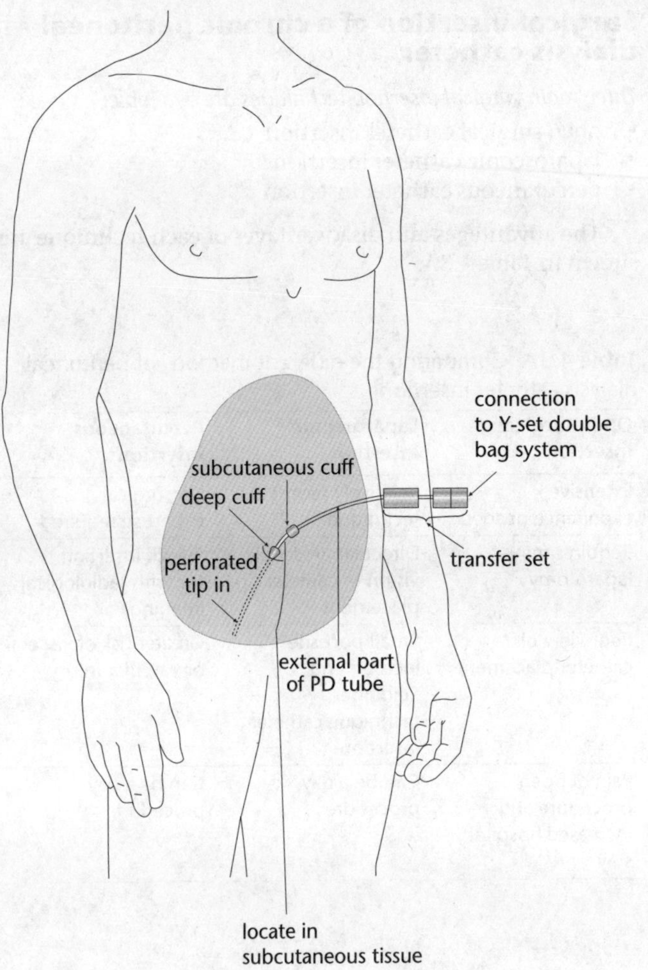

Fig. 4.3 The peritoneal dialysis catheter

Surgical insertion of a chronic peritoneal dialysis catheter

Three main surgical insertion techniques are available:

- open surgical catheter insertion
- laparoscopic catheter insertion
- percutaneous catheter insertion

The advantages and disadvantages of each technique are shown in Table 4.3A.

Table 4.3A Comparing the different methods of peritoneal dialysis catheter insertion

Open surgical insertion	Laparoscopic insertion	Percutaneous insertion
Extensive experience needed	Relatively recent technique	No surgical expertise required
Requires mini-laparotomy	Direct laparoscopic vision of catheter tip placement	'Blind' insertion (or with radiological imaging)
Poor view of catheter placement	Small port site incision and Seldinger technique catheter insertion	Greater risk of visceral or vascular injury
Pain of open procedure and increased hospital stay	Can be a day procedure	Can be a day procedure

Catheter position

- Approximately 15 cm of the catheter is within the peritoneal cavity with the tip placed into the pelvis. The tip may or may not be sutured.
- The catheter is tunnelled subcutaneously and brought out in the left iliac fossa (to avoid subsequent right side renal transplant).
- The external part of the catheter runs inferior to the exit site.

Guidelines for the placement and care of the chronic peritoneal dialysis catheter

The guidelines for the placement and care of the PD catheter are given in Protocol 4.3A.

Acute complications of insertion

- Haemorrhage.
- Haematoma.
- Visceral perforation.
- Malposition/failure of drainage.

Subacute–chronic mechanical complications of peritoneal dialysis catheter insertion

- Drainage problems (i.e. catheter not draining or infusing slowly, see Protocol 4.3B).
- Pain on infusion (see Protocol 4.3C).
- Complications due to increased abdominal pressure (e.g. dialysate leak, hernia, back pain) (see Protocols 4.3D and 4.3E). (Risk factors: age, polycystic kidney disease and high BMI.)

PROTOCOL 4.3A

How to place and care for the peritoneal dialysis catheter

Preoperative assessment

Referral to a renal physician and PD unit

Patients should be referred as early as possible for assessment of suitability for PD and provision of adequate education.

Surgical referral

Patients should be referred to a surgeon who is experienced in PD catheter placement at least 6 weeks before dialysis is needed. Consideration should be given to simultaneous placement of a native AV fistula (see Chapter 3.4).

Perioperative and intraoperative management

Duration of hospital admission

- 1 to 3 days.

Preoperative preparation

- Bowel preparation (i.e. two Coloxyl with senna in the evening for 2 days prior to surgery).
- Routine preoperative investigations: biochemistry, FBC, ECG, CXR.
- Marking the exit site. It is recommended that the exit-site position is marked on the patient's abdomen by the PD nurse (generally the night before catheter insertion), taking into account the belt line and skin folds.
- Other: shower with antiseptic soap solution on the day prior to and morning of surgery. NBM after midnight.
- Preoperative antibiotics: IV cephalosporin (e.g. cephazolin 1 g IV) at the time of anaesthetic induction.

Postoperative care

Exit-site dressing

- An adhesive dressing is left intact until outpatient review, 1 week later in the PD unit

Care of the catheter

- Patients should be instructed to anchor the catheter with micropore tape, as trauma to the catheter will increase the healing time.

Bathing/showering/swimming

- Patients are advised not to take a shower or bath or go swimming for at least 6 weeks after PD catheter insertion. Sponge baths are allowed. Thereafter, patients are allowed to shower. Swimming should be avoided until the exit site has healed completely. If swimming is undertaken, public pools should be avoided. Bathing in private pools and the beach are allowed, but a watertight dressing should be used. Baths should be avoided at all times due to the risk of bacterial contamination.
- Patients are advised to maintain regular bowel habits. This may require the use of regular aperients.

Follow-up after discharge from hospital

- Patients are reviewed at least weekly by the PD unit following discharge from hospital. The exit site is checked and the non-occlusive dressing is changed at each review.
- At least 14 days post-insertion is required before a patient commences training for PD in order to allow fibrous tissue to seal the PD tunnel.

PROTOCOL 4.3B

How to assess and manage peritoneal dialysis catheter drainage problems

Clinical presentation

- Catheter not draining and/or infusing (or very slow).

Causes

- Constipation (stool-filled bowel occluding the drainage holes of the catheter).
- Obstruction (kinked tube, internal blockage with fibrin, blood or omental wrap).
- Catheter migration out of pelvis.
- PD fluid loculated in an isolated area of the peritoneal cavity.

Management

- Try simple methods (e.g. change body position, treat constipation).
- If no improvement with simple measures:
    - check catheter position on plain abdominal X-ray (catheter tip should be located in the pelvis)
    - assess need for urgent HD (check biochemistry, fluid status)
    - consult surgeon to remedy underlying problem (see Table 4.3B)

Table 4.3B Surgical management of peritoneal dialysis catheter complications

Complication	Surgical management
Catheter tip migration	Laparoscopic repositioning ± suturing catheter tip to maintain position in pelvis
Catheter kinking	Kinking of catheter in early post-operative period may be amenable to repositioning but may require replacement
Catheter occlusion (adhesions, omental wrap)	Adhesions may be divided laparoscopically. Omentum may be excised via an open or laparoscopic approach
Visceral pain secondary to position	May be relieved by repositioning

PROTOCOL 4.3C

How to manage pain when infusing peritoneal dialysis fluid

Clinical presentation

- Pain occurs during the first few minutes of intraperitoneal infusion and then gradually subsides.
- Pain may occur transiently in new patients but is rarely a long-term problem.

Causes

- Acidity (pH 5.2–5.5) of the conventional lactate-buffered PD solutions.
- Catheter position.
- PD solution temperature.
- Hypertonicity of glucose solutions.

Management

- Exclude peritonitis (see Chapter 4.8).
- Try simple measures: reduce infusion rate, avoid hypertonic solutions.
- If persistent:
  - check for catheter complications (as in Protocol 4.3B)
  - consider using PD solutions with a neutral pH (Physioneal, Balance) (see Chapter 4.5)

PROTOCOL 4.3D

How to assess and manage a suspected peritoneal dialysate leak

Clinical presentation

- Oedema of the abdominal wall, perineum, scrotum or penis.
- External fluid at exit site tests positively for glucose (to distinguish from exit-site exudate).
- Unilateral pleural effusion (usually right-sided).
- Reduced drainage volumes.

Mechanisms

- Tear or anatomical defect in the peritoneal cavity so that dialysate leaks into the subcutaneous tissue (exacerbated by immediate use, delayed healing, hernia or congenital abnormality).

Management

- Stop PD for 1–3 weeks to allow tissue to heal.
- Use temporary HD support *or* low-volume (500–1500 mL) PD exchanges with cycler (see Chapter 4.6).
- Consult the surgeon. The cause of a persistent leak may need surgical correction (e.g. hernia) or the patient may need permanent transfer to HD. Pericatheter leaks may resolve with catheter replacement.
- Avoid activities that increase intra-abdominal pressure.
- A technetium PD scan may be useful if the anatomical site of the leak is not clear.

PROTOCOL 4.3E

How to manage a hernia/prolapse in a peritoneal dialysis patient

Clinical signs

- Sites: inguinal, incisional, umbilical, ventral hernia; vaginal prolapse.
- Look for complications (e.g. leakage, incarceration).

Mechanism

- Increased intra-abdominal pressure with a congenital abnormality or a new defect.

Management

- Stop PD.
- Consult surgeon for surgical repair.
- Use HD support for 4 weeks post-repair; or no dialysis for a few days followed by low-volume PD exchanges with a cycler (if good residual renal function).

Reference

1. Kelly J, McNamara K, May S. Peritoneoscopic peritoneal dialysis catheter insertion. Nephrology 2003; 8: 315–17.

Chapter 4.4
Peritoneal dialysis training and education

HIGHLIGHT

- Following catheter insertion, patients require close follow-up by the PD unit for education and training.

Commencement of peritoneal dialysis following insertion

- After surgical insertion, the PD catheter is not used for 10–14 days in order to allow healing of the insertion site and let a fibrous tissue seal form around the dacron cuffs.
- This period can be extended to 3 months, depending on how quickly the patient needs to start dialysis.
- Occasionally, it is necessary to use a PD catheter immediately. In this case, low-volume PD can be initiated (<1 L exchange volumes), either manually or using a cycler, after consultation with the surgeon (see Chapter 4.6).

Peritoneal dialysis training

- This is performed on an outpatient basis in the PD unit of the parent hospital by specialist PD nurses.
- Training usually takes 1–2 weeks.
- Patients or their carers are educated about:
  - the technique of PD exchange
  - PD solutions and equipment
  - exit-site and chronic catheter care

  - the use of antibiotics for peritonitis and exit-site infection
  - other common problems
- Patients are instructed to keep daily records of their body weight, BP, the number of PD exchanges, tonicity of the PD solution and UF volumes. The record book should be reviewed during all follow-up visits at the PD unit.

Problems during training

- Training is often difficult in patients with language difficulties or with other significant medical problems (e.g. severe uraemia, visual difficulties, debilitation).
- Under these circumstances, a member of the patient's family or a carer may need to be trained, or temporary HD provided to eliminate uraemic symptoms.

Chronic care of the peritoneal dialysis catheter

- Patients are educated about the chronic care of the catheter and exit site. This is essential in order to prevent trauma, exit-site infections and peritonitis.
- Chronic catheter care includes:
  - assessment of the exit site (i.e. for colour, presence of discharge, crust formation)
  - cleaning and dressing of the exit site (e.g. non-occlusive dressing, such as Cutiplast)
  - immobilising the catheter
  - protection from trauma
- Patients are advised not to swim in public pools and to avoid baths. Showers are permitted, as is swimming at the beach and in a private pool, providing a watertight dressing is used.

Chapter 4.5
Peritoneal dialysis solutions and connection systems

HIGHLIGHTS

- Dextrose-containing solutions are the standard solution.
- Newer solutions have some advantages but the long-term benefits are not proven.
- Connection systems are based on the Y-connection principle.

Peritoneal dialysis solutions

- Several different types of PD solutions are available (see Table 4.5A).
- PD solutions vary according to the:
  - *osmotic agent*: dextrose (1.5%, 2.5%, 4.25%), icodextrin (7.5%) or amino acids
  - *volume of PD solution*: 1, 1.5, 2, 2.5 or 3 L (5 L for use with a cycler only)
  - *ionised calcium level*: 1, 1.25 or 1.75 mmol/L
  - *buffer* and *pH*: lactate (35 mEq/L) or bicarbonate/lactate (10–15/25 mEq/L)
  - *bag configuration*: single, dual or triple chamber
- The standard electrolyte composition of most solutions is: Na (132 mEq/L), Cl (102 mEq/L), Mg (0.5 mEq/L) and K (0 mEq/L).

Table 4.5A Peritoneal dialysis solutions currently available in Australia

PD solution	Osmotic agent	Osmolality (mmol/kg)	pH	Buffer	GDP[(a)]	No. of chambers
Standard						
Dianeal	Dextrose (1.5–4.25 g/dL)	358–511	5.2–5.5	Lactate	35–40 +++	Single
Andy-disc or Stay-safe						
New						
Extraneal	7.5% Icodextrin	284	5.8	Lactate	+	Single
Nutrineal	1% amino acids	365	6.7	Lactate	+	Single
Physioneal	Glucose (1.5–4.25 g/dL)	358–511	7.4	Bicarbonate	+	Dual
Balance	Glucose (1.5–4.25 g/dL)	358–511	7.4	Lactate	+	Dual
Gambrosol Trio	Glucose (1.5–4.25 g/dL)	358–511	7.0	Lactate	+	Triple

(*a*) Glucose degradation products are formed during the manufacturing processing. GDPs may lead to the formation of advanced glycation end-products (AGE) in the peritoneum and *may* contribute to long-term peritoneal membrane failure.

Dextrose-containing peritoneal dialysis solutions

The standard PD solution uses dextrose as an osmotic agent. The higher the dextrose content (or tonicity), the more water is removed into the peritoneal cavity:

- 1.5%—removes a small amount of fluid (e.g. 100–200 mL)
- 2.5%—removes a moderate amount of fluid (e.g. 200–500 mL)
- 4.25%—removes a large amount of fluid (e.g. 500–1000 mL)

Although dextrose-based solutions are currently the mainstay of PD therapy and have been used for more than two decades, they are associated with problems (see Table 4.5B) due to their:

- high glucose content (which has local and systemic effects)
- non-physiological pH
- hyperosmolality
- propensity to form GDPs (see Tables 4.5A and 4.5B)

Table 4.5B Advantages and disadvantages of dextrose peritoneal dialysis solutions

Advantages	Disadvantages
Caloric load (weight gain) if malnourished	Caloric load (weight gain)
Long-term safety established	Glucose loading (dyslipidaemia/ ?atherogenic, diabetogenic) Glucose-induced peritoneal fibrosis Generation of GDPs Non-physiological pH and osmolality Reduction of peritoneal defences (?increased risk of peritonitis)

New peritoneal dialysis solutions

- These have been developed due to the adverse effects of dextrose-based PD solutions.

- They are more biocompatible than dextrose-based PD solutions but are more expensive.
- The long-term influence on technique survival is unproven.
- Their use is currently limited to specific clinical situations.

Types and use of new peritoneal dialysis solutions

Icodextrin (Extraneal)

What is it?

- An isoosmolar mixture of high-MW, soluble, polymaltose glucose polymers, with limited diffusive capacity across the peritoneal membrane. It provides sustained *colloid-based* UF over long dwell times. Some absorption of icodextrin occurs, which is metabolised to maltose.

Clinical evidence

- Icodextrin is equivalent to 4.25% dextrose solution in terms of UF capacity and solute clearance, and improves dyslipidaemia. No long-term randomised clinical trials have shown that it improves technique survival and/or reduces cardiovascular morbidity.

How to use

- Once daily dwelling for a long-period (e.g. overnight 8–10 hours). Use more than once per day is not recommended due to the lack of long-term safety data, particularly about the effects of maltose exposure.

Side-effects

- The side-effects of maltose absorption include:
  - under-estimation of blood glucose level when using glucometers based on glucose dehydrogenase method (e.g. Advantage, Accucheck, Glucotrend)
  - under-estimation (up to 90%) of serum amylase level (i.e. falsely low readings)
  - other long-term effects unknown

- Hypersensitivity: skin rash and sterile peritonitis (rare).
- Mild hyponatraemia.

Indications for use

- Patients who develop refractory fluid overload using dextrose-based solutions (UF failure). In these patients icodextrin may delay or prevent transfer to HD.

Amino acids (Nutrineal)

What is it?

- Amino acids (87 mmol/L, 61% essential) are used as the osmotic agent (2.0 or 2.5 L). The advantage is that the absorption of the amino acids (18 g, about 25% of a PD patient's required protein intake; 1.2 g/kg/d) can be used to treat concurrent malnutrition in PD patients.

Clinical evidence

- Several studies show that nutritional indices can be improved but there are no long-term data on the effects on patient and technique survival. The effects on residual renal function are unknown.

How to use

- A single exchange coinciding with the major meal of the day (e.g. lunch, dinner).

Side-effects

- Dose-dependent nausea, vomiting, uraemic symptoms and mild metabolic acidosis.

Indications for use

- Short-term use in severely malnourished PD patients, in conjunction with dietary intervention.

Bicarbonate/lactate-buffered peritoneal dialysis solutions (Physioneal, Balance)

What is it?

- Lactate is the traditional buffer in conventional solutions but the pH achieved is unphysiological

(5.5–6.5) and may possibly contribute to pain on infusion of the PD fluid and formation of GDPs (peritoneal fibrosis). Bicarbonate-buffered solutions have a reduced lactate content, giving a more physiological pH (7.0–7.6). The solutions are provided in dual-chamber bags that are mixed immediately prior to use.

Clinical evidence

- Studies show that these solutions may be more biocompatible, reduce infusion pain and possibly reduce peritonitis. No long-term clinical studies are available.

How to use it

- 4–5 times daily, as per standard regimens.

Side-effects

- As for dextrose-containing PD solutions.

Indications for use

- Cost is prohibitive for use in all patients. Until further evidence about the long-term benefit is available, this solution should be considered. They may be indicated in selected patients in whom peritoneal membrane preservation is important (e.g. patients in whom transfer to HD is not feasible) or in those with intractable pain on infusion.

Reduced GDP content (Gambrosol Trio)

What is it?

- The PD bag has three compartments, in which glucose is kept separate from the buffer and electrolytes, thereby reducing GDP formation. Each compartment contains a different dextrose concentration. Therefore, any of the three dextrose concentrations can be formed from a single PD bag.

Clinical evidence

- Studies show that the solution is more biocompatible but no long-term data are available.

How to use it and indications for use

- As for bicarbonate-buffered solutions. Because any combination of dextrose concentrations can be formed from a single bag, there is the added benefit of reduced storage space.

Connection systems

PD solutions come in PD bags that are usually attached to a Y-set double bag system. The Y-system refers to the configuration of the connecting tubes (see Figs 4.1 and 4.5, Table 4.5C)—the stem is connected to the PD catheter and transfer set, and the afferent and efferent limbs connect to a fresh dialysate PD bag and empty drainage bag respectively. A clamp controls the rate and direction of fluid flow, either drainage or infusion. Y-connection systems have dramatically reduced the incidence of peritonitis because an initial priming with the PD solution and drainage of the main stem allows bacteria present in the main stem to be washed into the drainage bag, rather than enter the peritoneum.

Delivery and storage of peritoneal dialysis solutions

- Solutions are delivered to the patient's home by the manufacturer every month.

Table 4.5C Connection systems

Manufacturer	Type	Description
Baxter	Freeline solo	Y-connector system with clamps
Fresenius	Stay-safe	Y-connector systems with special disc and pins, instead of clamps. May be useful in patients with poor vision.
	Andy-disc	
Gambro	Trio	Y-connector system with clamps

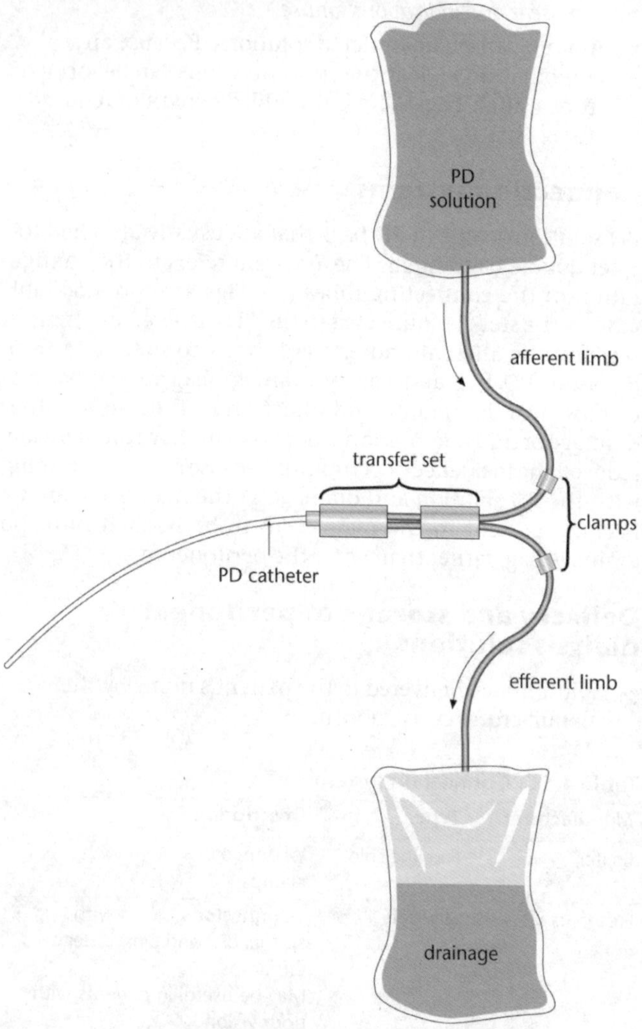

Fig. 4.5 Principle of the Y-connection system

- A 1-month supply requires a large storage area (2 m^3).
- PD solutions can be delivered to almost any geographic location in Australia or overseas, making it easy for PD patients to travel.

References

1. Johnson DW, Agar J, Collins J, et al. Recommendations for the use of icodextrin in peritoneal dialysis patients. Nephrology 2003; 8: 1–7.
2. Vardhan A, Zweers MM, Gokal R, Frediet RT. A solutions portfolio approach to peritoneal dialysis. Kidney Int 2003; 64: S114–S123.

Chapter 4.6
Methods of chronic peritoneal dialysis

HIGHLIGHTS

- Continuous ambulatory peritoneal dialysis (CAPD) and automated peritoneal dialysis (APD) are the two most common methods of chronic PD.
- In APD, a machine (cycler) performs the exchanges (usually overnight while the patient is asleep).

Definition of a peritoneal dialysis exchange and cycle

Peritoneal dialysis exchange

This is the basic procedure of PD. An exchange is defined as the *drainage* of old solution from the peritoneal cavity and *infusion* of a specific volume of fresh PD solution through a Y-connection system attached to the PD catheter/ transfer set.

Drainage

Each exchange begins with drainage of the PD solution remaining from a previous cycle. Depending on the original volume instilled, drainage can take about 20 minutes.

Infusion

PD solution flows into the peritoneal cavity by gravity through a connection system, which usually takes 10 minutes. Factors that can affect the rate of inflow include height of the solution, inner diameter of the tubing, tubing length and intra-abdominal pressure. Each exchange can therefore take about 30 minutes.

Peritoneal dialysis cycle

The PD cycle is the period of time from the beginning of one exchange until the next, and includes the drainage and infusion time, as well as the *dwell* time.

Dwell

Following infusion, the PD solution is allowed to dwell in the peritoneal cavity for diffusion and UF to take place. The dwell time depends on the clinical situation and type of PD.

Continuous ambulatory peritoneal dialysis

- This is the most common form of PD.
- Most patients perform 4 bag exchanges per day (3 cycles during the day of 4–6 hours each and one cycle at night of 8–10 hours). The exchanges should be spaced throughout the day, with the last exchange taking place prior to bedtime.
- The dialysis is continuous, 24 hours a day, each day of the week.
- The exchanges can be performed in any clean and convenient environment.
- Following each exchange the patient is free to conduct his or her normal daily activities, and hence the concept of the PD being *ambulatory*.

Automated peritoneal dialysis

- In APD, a small portable machine, called a cycler, performs the exchanges of PD solution automatically. Usually the patient connects to the cycler each night before going to sleep, and the cycler performs the exchanges overnight.
- A variety of cyclers are available from different manufacturers (see Table 4.6).
- The exchanges comprise frequent short dwells (e.g. 4 or more exchanges of 4 × 2 L over a 9–10 hour period), with the last one ending with a 'fill' (e.g. 2 L), leaving the patient with a 'wet' day. Often the patient will perform

Table 4.6 Cyclers available in Australia

Baxter	Fresenius	Gambro
Homechoice	Sleep-safe	Serena
Quantum[(a)]		

(*a*) Provides a single extra exchange at night.

another exchange at some time during the day. In some circumstances, the patients may have a 'dry' day (e.g. post-hernia repair) (see Chapter 4.3). The cycler calculates the amount of fluid infused and the drain volumes.

- In comparison to CAPD, the total dialysate volumes are larger for APD.

When to use

- If there is failure to achieve clearance targets with CAPD.
- The patient has a high transporter status on the PET (due to UF failure as a result of rapid solute absorption).
- In the presence of hernias, dialysate leaks, prolapse, divarication of rectus abdominis.
- For lifestyle and social reasons (e.g. working during the day, elderly with carers).
- In children.

Chapter 4.7
Prescribing and modifying chronic peritoneal dialysis

HIGHLIGHT

- Prescription of PD for a patient must be reviewed on a regular basis and includes an assessment of dialysis adequacy and peritoneal membrane function.

Deciding on the initial method of chronic peritoneal dialysis

- CAPD is the usual initial method of chronic PD.
- APD has specific indications (see Chapter 4.6).

Initial prescription in continuous ambulatory peritoneal dialysis

The initial CAPD prescription must specify the following (see Protocol 4.7):

- frequency of exchanges
- volume of exchanges (depends on patient size and residual renal function)
- dextrose concentration of PD solution (depends on the UF requirements)

Dextrose solutions are the first choice until the long-term benefits of new PD solutions are known from ongoing clinical trials.

PROTOCOL 4.7

How to write the initial chronic continuous ambulatory peritoneal dialysis prescription

Specify the number of exchanges per day

- Three during the day (every 6 hours) and 1 overnight (8–10 hours).
- Increase (up to 5 during the day) if the patient has fluid overload or uraemia (increases solute removal).

Decide what volume of peritoneal dialysis solution to use

This depends on the size of the patient (thus, calculate BSA[a]), residual renal function (which determines the low-MW solute removal and is assessed by urine output) and timing after PD catheter insertion:

- if BSA < 2.0 m^2, then 4×2.0-L exchanges
- if BSA > 2.0 m^2, then 4×2.5-L exchanges
- increase this by 0.5 L if patient has poor residual renal function (urine output $<$200 mL/d)
- reduce by 0.5–1 L if CAPD begins within 2 weeks of catheter insertion

Determine the dextrose concentration of the peritoneal dialysis solutions

The concentration of the PD solution depends upon the degree of residual renal function and the effectiveness of fluid removal (UF) with different dextrose concentrations (this varies from patient to patient). In general, 1–2 L of total UF (dialysis + residual renal) daily is necessary to maintain satisfactory fluid balance and solute clearance. Further UF may be needed depending on the fluid status of the patient:

- no extra UF requirements (e.g. $>$500 mL urine output, not at risk of fluid overload)—use $4 \times 1.5\%$ dextrose bags
- extra UF needed (e.g. no significant urine output or at risk of fluid overload)—alternate 1.5% with 2.5% dextrose bags (use 1.5% as the overnight bag)

- avoid use of 4.25% dextrose bags unless severe gross fluid overload (the excessive exposure has the potential to damage peritoneum in the long term)

(*a*) Estimate the BSA using a standard nomogram chart (using patient's height and weight, see Table 4.7A) or calculate the BSA directly using the DuBois formula: BSA = 0.007 184 × (patient height, cm)$^{0.725}$ × (patient weight, kg)$^{0.425}$.

Modifying the prescription after the first month

- This depends on the clinical assessment: uraemic symptoms, fluid status and results of the PET at 6 weeks (see Table 4.7B).

Modifying the prescription at 6–12 months

- In addition to parameters assessed after the first month, closer attention is paid to clinical and biochemical parameters of dialysis adequacy (see below).

Adequacy of dialysis in peritoneal dialysis

Definition of dialysis 'adequacy'

- Adequate dialysis—the minimum effective dose.
- Optimal dialysis—the dialysis dose above which the incremental clinical benefit does not justify patient burden or financial costs.

Assessment of dialysis adequacy

Dialysis adequacy is routinely performed on a 6-monthly basis or when considered necessary. To assess the adequacy, a number of parameters must be taken into consideration:

- clinical factors (nutrition, well-being, uraemic symptoms, quality of life)
- residual renal function
- blood results (urea, creatinine, albumin)
- biochemical markers of dialysis adequacy (Kt/V, creatinine clearance)

Table 4.7A Estimating body surface area

Weight (kg)	Height (cm)								
	120	130	140	150	160	170	180	190	200
40	1.11	1.17	1.24	1.30	1.37	1.43	1.49	1.55	1.61
50	1.22	1.29	1.36	1.43	1.50	1.57	1.63	1.70	1.76
60	1.32	1.40	1.47	1.55	1.62	1.69	1.77	1.84	1.91
70	1.41	1.49	1.57	1.65	1.73	1.81	1.89	1.96	2.04
80	1.49	1.58	1.66	1.75	1.83	1.92	2.00	2.08	2.15
90	1.56	1.66	1.75	1.84	1.93	2.01	2.10	2.18	2.27
100	1.64	1.73	1.83	1.92	2.02	2.11	2.20	2.28	2.37
110	1.70	1.81	1.91	2.00	2.10	2.19	2.29	2.38	2.47

Table 4.7B Modifying the peritoneal dialysis prescription according to the peritoneal equilibration test[a]

Transporter type	Dialysis capacity (solute transport)	UF	Clinical problems	Proportion of PD patients	PD prescription
High	Very good (highly efficient)	Poor	Increased glucose absorption Hypoalbuminaemia	1/3	Best suited for short frequent cycles, especially (APD), *or* standard CAPD plus icodextrin
High-average	Good (efficient)	Good	None	1/3	Standard CAPD
Low-average	Good to poor (less efficient)	Good to excellent	None	1/3	Standard CAPD
Low	Poor (inefficient)	Excellent	Inadequate dialysis if no residual renal function	<5%	PD may not provide adequate dialysis if there is little residual renal function

(*a*) Membrane characteristics vary with time and acute complications (e.g. patients with acute peritonitis develop membranes with high transporter characteristics possibly due to increased vascular permeability).

Biochemical assessment of dialysis adequacy

Measurement of dialysate *Kt/V* (urea clearance) and creatinine clearance is based on the volume of dialysate effluent and its urea and creatinine concentration over a 24-hour period. To this must be added the residual renal function. There is no consensus on which solute (urea or creatinine) is the best to measure. In reality, neither is ideal as they do not reflect clearance of larger solutes.

Biochemical target of dialysis adequacy

The CARI guidelines state that CAPD and APD patients should achieve a weekly *Kt/V* of 1.7–2.0 and a corrected creatinine clearance of 60 L/week (high transporters) or 50 L/week (low transporters).

However, in light of the recent Adequacy of Peritoneal Dialysis in Mexico (ADEMEX) randomised control trial, if a PD patient is *doing well clinically* but does not meet clearance targets, there is no evidence of increased mortality when peritoneal *Kt/V* is at or above 1.7 and peritoneal creatinine clearance is at or above 45–50 L/week. The quality of life and life expectancy of the patient as well as patient preference also influence modifications to the treatment regimen.

Methods to increase dialysis dose in order to increase solute clearance

- Minimise loss of residual renal function (see below).
- Increase the fill volume (the most effective method), which thus increases the concentration gradients for solute removal. Methods to promote this in the patient include:
    - using a large volume of PD solution at the time of least activity (e.g. overnight bag)
    - using a lower percentage of glucose with larger volumes
    - changing the fill volume gradually by 0.5-L increments
    - giving education and giving positive reinforcement to patients

- Increase the number of exchanges (from 4 to 5 per day). The Quantum device allows an extra overnight exchange (see Chapter 4.5) in PD patients who have difficulty performing an extra daytime exchange.
- Consider using overnight APD with CAPD.

However, there are several problems with all of these strategies. For example, patients may not tolerate increasing the PD bag volume or may not be willing to perform 5 exchanges and/or use a cycler. In these situations, the only options are either transfer to HD or to continue until the patient agrees to transfer to HD.

Residual renal function

- This declines once a patient commences dialysis (less quickly in patients on PD than on HD).
- In PD patients, patient survival and adequate dialysis are dependent on residual renal function.
- Therefore, even after dialysis commencement, residual renal function should be protected to preserve, for as long as possible, clearance of water, small solutes and middle molecules and endocrine functions of the kidney (e.g. erythropoietin production, vitamin D activation).
- Absent residual renal function is defined as a urine volume < 200 mL per day.
- Risk factors for loss of residual renal function include obesity, high transporter status, high dietary intake of protein and exposure to nephrotoxins.

Protecting residual renal function

- Avoid nephrotoxins (e.g. aminoglycosides, NSAIDs and coxibs).
- Administration of an ACEI may slow the loss of residual renal function.
- Frusemide (up to 250–500 mg/d) may preserve water clearance and limit the use of high concentrations of glucose in PD solutions but it will not alter the decline in residual renal function.

Patients who are anuric or have a large muscle mass

In patients with little or no residual renal function or large muscle mass, it may be difficult to provide adequate dialysis with conventional PD (especially those with low transporter status), and overnight APD with CAPD or transfer to HD is often required to provide adequate dialysis.

Prescribing automated peritoneal dialysis

- APD is best suited to patients with high-transporter type membranes. Other indications are listed in Chapter 4.6.
- Typical APD regimen: 10 L (2 × 5.0-L bag) over 8–10 hours (3–5 exchanges of 2.0–3.0 L each) with or without 1–2 long day dwells (1.5–2.5 L). Icodextrin-containing dialysate could be used for the long day dwell.
- The prescription is entered electronically into the cycler.
- The higher the membrane transport status, the more frequent the exchanges to be given.
- The larger the patient, the greater the required fill volume.
- Anuric patients nearly always require a daytime exchange.
- The dialysis prescription is easier to modify because the required number of exchanges is determined and performed by the machine, and changes in volume are tolerated better when the patient is supine. On the other hand, the limitations are connection time and the cost of using larger volumes of dialysate (>10 L).
- Patients with large BSA or anuric patients with a low transporter membrane may not achieve adequate dialysis even with APD, as for CAPD.

References

1. Paniagua R, Amato D, Vonesh E, et al. Effects of increased peritoneal clearances on mortality rates in peritoneal dialysis: ADEMEX, a prospective, randomized, controlled trial. J Am Soc Nephrol 2002; 13: 1307–20.

Chapter 4.8
Peritonitis and exit-site infections

HIGHLIGHTS

- Early recognition and treatment of peritonitis and exit-site infection is essential.
- Inadequate or delayed treatment of peritonitis can lead to PD catheter loss and transfer to HD or, rarely, death.

Peritoneal-dialysis-related peritonitis

Definition and overview

- This is acute inflammation of the peritoneum in a chronic PD patient, usually due to bacteria.
- It is a common problem and, on average, a patient may experience one episode every 18–24 months of treatment time.
- It usually arises from contamination of the peritoneum by the patient's own flora, either due to:
  - *transluminal* contamination (most common)—organisms enter the peritoneum through the catheter due to touch or aerosol contamination of the PD fluid or equipment as a result of a breach in sterile technique
  - *endogenous* contamination—it is presumed that organisms arise from transmural migration across the bowel wall (e.g. constipation, diverticular disease), or
  - *exogenous* contamination—organisms enter via an exit-site infection

- Gram-positive organisms (*Staphylococcus epidermis* and *Staphylococcus aureus*) are the most frequent pathogens and cause over 70% of peritonitis episodes. Other causes include Gram-negative organisms (*Escherichia coli, Klebsiella, Pseudomonas*), and 'culture'-negative peritonitis is not uncommon. MRSA remains an uncommon cause of peritonitis.
- *S. aureus,* Gram-negative and fungal peritonitis are more likely to be associated with slower treatment response, septicaemia, high mortality and/or catheter loss and transfer to HD.
- Typically, only one organism is isolated on culture, and multi-Gram-negative peritonitis suggests perforated bowel or other intra-abdominal pathology, requiring urgent surgical review.
- Patients respond rapidly to antimicrobial therapy. Early surgical review and catheter removal should be undertaken in those who fail to improve within 72 hours and those with multi-Gram-negative or fungal peritonitis in order to prevent mortality and irreversible scarring of the peritoneum.
- Vancomycin should be avoided as the initial antibiotic therapy because of the emergence of resistant strains of enterococci (VRE). Vancomycin should be reserved for the treatment of methicillin-resistant *S. epidermis* (MRSE) or MRSA.

Diagnosis

Clinical suspicion

- Cloudy effluent.
- Abdominal pain with or without guarding.
- Nausea/vomiting.
- Poor drainage.
- Diarrhoea/constipation.

Confirmation of diagnosis with urgent microscopy and culture

- Dialysate WBC count $>100 \times 10^6$/L with $>50\%$ neutrophils.
- Positive culture.

Management

Initial and subsequent management is described in Protocols 4.8A and 4.8B, and modification of the antibiotic regimen in Tables 4.8A–4.8C.

Exit-site infection

Definition and overview

- This is an inflammation of the catheter exit site with purulent drainage and positive culture. *Note:* Positive culture in the absence of inflammation does not indicate infection because virtually all healed exit sites are colonised by bacteria.
- *S. aureus* is the most common causative organism followed by polymicrobial Gram-positive cocci, Gram-negative and rarely fungal infection.

Diagnosis

- Surrounding skin is red (>13 mm or 2× catheter diameter; equivocal if <13 mm).
- Swelling.
- Tenderness or pain.
- Discharge (purulent and/or crust formation).
- Swab showing neutrophils with positive culture.

Risk factors

- No prophylactic antibiotics given at catheter insertion.
- Upward directed tunnel and exit site.
- Dialysate leak or frank bleed.
- Delayed or ineffective exit-site healing, skin breakdown.
- Poor hygiene, sweating, trauma at the catheter.

Management

Protocol 4.8C shows how to manage exit-site infections.

PROTOCOL 4.8A

How to initially manage suspected peritoneal-dialysis-related peritonitis (0–48 h)

- Send PD fluid to laboratory for urgent WBC count and culture.
  - Withdraw two 20-mL samples from the effluent bag, using a sterile technique.
  - Use one sample (2 × 10 mL) to inoculate a set of blood culture bottles.
  - Inject the other 20-mL sample into a sterile container. This is used for the WBC count and quantitative culture.
- Commence empirical antibiotic therapy immediately.
  - Use the IP route for antibiotics in the same PD bag for 6–8 h dwell.
  - Use the IV route if the patient has systemic sepsis.

Antibiotic therapy for suspected PD-related peritonitis

Agent	Little RRF (UO <500 mL/d)	RRF (UO >500 mL/d)
Cephazolin	1 g daily	Increase dose by 25%
Gentamicin	80 mg daily for 2 consecutive days (or 60 mg if BW <40 kg)	80 mg for 2 consecutive days (or 60 mg if BW <40 kg)

- Substitute cephazolin with vancomycin (2 g IP every 3–5 days depending on trough levels in blood) if:
  - history of MRSA infection or carriage
  - recent or current exit site infection
  - features of severe infection (e.g. fever severe abdominal pain).
- Decide if the patient needs hospitalisation.
  - Admit the patient if systemically unwell (temperature >39°C), has haemodynamic compromise or in severe pain.
  - If no hospital admission is required, arrange for daily follow-up by the PD unit.
  - Convert APD patients to CAPD during antibiotic treatment.

- Consider differential diagnosis of a 'cloudy' bag:
  - infective peritonitis
  - chemical or eosinophilic peritonitis (e.g. due to icodextrin)
  - intra-abdominal or gynaecological pathology
  - menstruation
  - fibrin
- Review progress daily.

PROTOCOL 4.8B

How to subsequently manage peritoneal-dialysis-related peritonitis (>48 h–3 weeks)

- Check culture results and modify antibiotic regimen according to organism (Tables 4.8A–4.8C).
- Consider the pathogenesis to prevent another episode (Table 4.8D).
- Evaluate and classify response to therapy:
  - *responsive:* symptoms decrease rapidly and resolve in 48–72 hours
  - *refractory:* symptoms do not abate after 72 hours of appropriate therapy
  - *relapsing:* recurs within 4 weeks of completion of antibiotic therapy, with organism and antibiotic sensitivities identical to the initial infection
- Consider indications for catheter removal:
  - refractory peritonitis >3–5 days
  - tunnel infection
  - polymicrobial peritonitis, fungal or tuberculous peritonitis
- Look for complications of peritonitis:
  - abscess formation
  - abdominal adhesions
  - changes in dialysis kinetics: increased clearance, increased glucose absorption, increased protein losses, increased UF rate early in dwell (total volume of long dwell is decreased due to earlier and more rapid absorption of glucose)

PROTOCOL 4.8C

How to manage exit-site infections

Minor erythema only

- Withhold antibiotics, take swab and await culture result.
- If culture is negative and minor erythema persists, treat with dicloxacillin 1 g IP and gentamicin 80 mg IP for 48 hours, and commence dicloxacillin 500 mg qid orally. (If allergic to penicillin, use cephalexin [500 mg tds] or clindamycin [300 mg tds] [if immediate hypersensitivity allergic reaction to penicillin]). If previous MRSA, use vancomycin.

More severe infection

- Swab and give dicloxacillin 1 g IP and gentamicin 80 mg IP for 48 hours.
- Commence dicloxacillin 500 mg qid orally.
- At 48 hours switch to appropriate oral antibiotics based on culture results. Continue for 4 weeks.
- If initial response is slow (i.e. 1–2 weeks) or if pus is present at the start, then treatment should be continued for 6 weeks.
- If exit-site infection worsens, repeat parenteral treatment based on the culture result.

Antibiotic management

Antibiotic management must be accompanied by vigilant exit-site care.

- Dress exit site daily with Primapore, changing frequently to keep exit site dry at all times.

Complications

Watch for complications of exit-site infection (e.g. tunnel infection, peritonitis that may require catheter removal and/or transfer to HD).

Table 4.8A Modification of the antibiotic regimen in the presence of proven Gram-positive peritonitis

Organism	Management	Total duration
Staphylococcus epidermis (MRSE) or MRSA	• Cease gentamicin and cephazolin • Give vancomycin 30 mg/kg (max 2 g) IP (total 3 doses) • Check trough level and repeat dose immediately on day 5 • Repeat vancomycin dose based on trough level taken on day 5	2–3 weeks
Staphylococcus aureus	• Cease gentamicin and continue daily cephazolin (1 g) until day 7	3 weeks
	• Add oral dicloxacillin 500 mg qid on day 7 and continue until day 21. If allergic to penicillin, use cephalexin (500 mg tds) or clindamycin (300 mg tds) (if immediate hypersensitivity allergic reaction to penicillin)	
Enterococcus	• Cease cephazolin • Continue with daily gentamicin (measure trough level every 72 h) • Commence ampicillin 1 g IP daily for 7 days • Commence oral amoxycillin 500 mg tds until day 14	2 weeks

Table 4.8B Modification of the antibiotic regimen in the presence of proven Gram-negative peritonitis

Organism	Management	Total duration
Single organism (non-*Pseudomonas*, *Xanthomonas*)	• Treat according to sensitivities with intermittent daily IP therapy for 7 days followed by oral antibiotics for another 7 days	2 weeks
Multi-Gram negative	• Treat according to sensitivities with daily IP therapy for 7 days • Add metronidazole 500 mg tds IV or 400 mg tds PO until day 21 • Consider bowel pathology and surgical intervention	3 weeks
Pseudomonas	• Cease cephazolin • Continue daily gentamicin (measure trough level every 72 h) for 7 days • Add ceftazidime 1 g daily and until day 7 • At least 2 days after clear bag (or day 7, whichever is later) switch to oral ciprofloxacin 250 mg tds for a further 14 days. (*Note:* Instruct patients to separate ciprofloxacin and phosphate binders by at least 2 hours, as the latter reduces ciprofloxacin absorption.) • With adequate response cease therapy on day 21 • Consider early catheter removal if inadequate response	3 weeks
Xanthomonas	• Cease cephazolin • Continue with daily gentamicin (measure trough level every 72 h) for 7 days • Add trimethoprim 320 mg/sulfamethoxazole 1600 mg daily for at least 21 days • Consider early catheter removal for inadequate response	3 weeks

Table 4.8C Modification of the antibiotic regimen in the presence of culture negative or fungal peritonitis

Organism	Management	Total duration
Culture negative	• *If clinical improvement at 48 hours,* cease gentamicin and continue cephazolin IP until day 7 • *If not improving at 48 hours,* repeat WBC count, Gram stain and culture • Continue empirical treatment until repeat culture available (4 days) • Consider early catheter removal if poor response	2 weeks
Fungal	• IP fluconazole (200 mg every 2 days) until early catheter removal • After catheter removal, oral fluconazole 200 mg/d for 10 days if tolerated and response to treatment rapid • If fluconazole is not tolerated and signs continue after catheter removal, then give IV amphotericin (increase to 30 mg/d until all signs are resolved), then oral fluconazole for 2 weeks. If fluconazole is not tolerated, continue IV amphotericin for 1 week after complete resolution of signs	2 weeks

Table 4.8D Prevention of peritonitis

Mechanism	Potential organism	Prevention
Contamination	• *Staphylococcus aureus, Staphylococcus epidermis* • Organisms found in the throat or water	Patient education
Tunnel infection	• *S. aureus, Pseudomonas*	Treat exit-site infections early and adequately
Haematogenous; transient with dental procedure, sigmoidoscopy	• *Streptococcus viridans*	Prophylactic antibiotics
Bowel wall: diverticular disease and severe vascular disease, unresolved constipation	• Enteric anaerobic organism	Treat constipation
Female reproductive tract	• Fungal	Treat infections; topical antifungal during antibiotic treatment if predisposed to candidiasis

Tunnel infection

This is infection of the subcutaneous tunnel between the two cuffs of the PD catheter. It is usually due to infection of the deep cuff. It is treated by antibiotics and can require catheter removal. Diagnosis may require ultrasound and/or CT scan.

Prevention of peritonitis and exit-site infection

- Good sterile technique and vigilant exit-site care.
- Delaying the first dressing change for 5 days, and start of showering and commencement of PD for 2 weeks.
- Intranasal mupirocin use for nasal carriers of *S. aureus*.
- Use of prophylactic antibiotics with invasive procedures (e.g. colonoscopy).

References

1. Thomas MC, Harris DCH. Management of bacterial peritonitis and exit site infection in continuous ambulatory peritoneal dialysis. Nephrology 2002; 7: 267–71.
2. Evidence for Peritonitis Treatment Working Group. CARI guidelines. Nephrology 2004; 9 (suppl 3): S41–S91.

Chapter 4.9
Medical complications of chronic peritoneal dialysis

HIGHLIGHTS

The medical complications of chronic PD include:

- problems related to inadequate PD dose
- fluid overload or volume depletion caused by disturbances of ECFV
- metabolic disturbances (e.g. hypokalaemia, malnutrition, dyslipidaemia, effects of glucose absorption)
- constipation
- social and psychological problems
- cardiopulmonary compromise

Inadequate peritoneal dialysis

Diagnosis

- Symptoms of uraemia.
- Malnutrition.
- Reduced *Kt/V* or creatinine clearance.

Note: A diagnosis of inadequate dialysis is a *clinical* diagnosis and not reliant entirely on laboratory measurements. See also Protocol 3.3A on assessing the adequacy of the HD dose.

Mechanisms

- Loss of residual renal function.
- Non-compliance.

- Peritoneal membrane failure. With time, patients on long-term PD develop submesothelial fibrosis and vasculopathy of the peritoneal capillaries. This can contribute to dialysis failure and may, in part, be related to bioincompatible glucose-containing PD solutions.
- Inappropriate conversion from APD to CAPD.

Management

The basic principles are the same as for management of inadequate HD (see Protocol 3.3B). Clinical methods to increase solute transport/clearance in PD include increasing the:

- volume of PD solution—increases concentration gradient
- frequency of PD exchanges—increases solute removal
- PD fluid tonicity—increases solute removal by solvent/water drag

  Other methods include:

- use icodextrin for long overnight dwell
- consider using APD + CAPD
- transfer to HD

Fluid overload and hypertension

Clinical signs

- Positive fluid balance.
- Weight gain (>2 kg above dry weight).
- Dyspnoea/oedema.
- HT.
- Raised JVP.
- Pulmonary oedema.
- Cardiac failure.

Mechanisms

- Inadequate fluid removal (e.g. use of inappropriately low dextrose PD solutions, malfunctioning catheter, reduced insensible losses due to reduced sweating in winter).

- Peritoneal membrane failure.
- Hypoalbuminaemia.
- Absorption of dialysate from overnight bag.
- Excess salt and water intake.
- Cardiac failure.

Management

- Reduce salt and fluid intake (500–800 mL/d).
- Monitor input and losses, and weigh daily.
- Increase glucose strength in bag and increase frequency of exchanges (e.g. change all PD bags to 2.5%, 5 exchanges per day).
- Consider frusemide 250–500 mg/d.
- Treat underlying cause of fluid overload (e.g. cardiac failure, ischaemic heart disease).
- Other: shorten dwell time, correct low serum albumin level.
- If refractory, consider use of icodextrin or temporary HD support for isolated UF.

Note:
- Chronic fluid overload is a strong predictor of poor outcome.
- HT in PD patients is predominantly volume (salt)-sensitive.
- Correction of volume expansion is associated with normalisation of BP and reduces the need for antihypertensives.

Volume depletion

Clinical signs

- Negative fluid balance.
- Weight loss (<1–2 kg below dry weight).
- Reduced skin turgor.
- Dry mucous membranes.
- Postural hypotension.
- Tachycardia.

Mechanisms

- Excessive fluid removal (e.g. inappropriate use of higher dextrose solutions).
- Extrarenal losses (increased insensible losses, gastrointestinal losses).
- Inadequate intake of salt and water.

Management

- Increase oral fluid intake by 1000 mL per day (consider IV fluid if severe).
- Decrease glucose strength of all PD bags to 1.5% ± increase dwell time.
- Review antihypertensive management.
- Treat underlying cause of volume depletion.

Metabolic abnormalities

Hypokalaemia

Clinical signs

- Weakness.
- Arrhythmias.
- Asymptomatic.

Mechanisms

In contrast to HD patients, PD patients may become hypokalaemic and up to 15% may require potassium supplementation. PD solutions contain no potassium and therefore potassium deficit may arise, especially if there are extrarenal losses (vomiting, diarrhoea) or inadequate dietary intake.

Management

- Increase dietary intake of potassium (e.g. bananas, stone fruit) and/or start potassium supplements (e.g. Slow K once bd).

Malnutrition

Clinical signs

- Weight loss.
- Reduced muscle mass.
- Hypoalbuminaemia.

Mechanisms

Patients on PD lose up to 15 g of protein per day (as protein, amino acids and other nutrients) and may not maintain adequate dietary intake. Consequently, up to 30% of PD patients may have evidence of mild–moderate malnutrition. Protein losses increase during peritonitis. Risk factors for malnutrition include low nutrient intake, impaired appetite, abdominal fullness, gastroparesis and inappropriate removal of metabolites. The clinical significance of malnutrition is that it is associated with an increased risk of morbidity and mortality.

Management

- Refer to dietitian.
- Oral nutritional supplementation.
- Temporary use of 1.1% amino acid PD solution with main meal—this may improve nutritional parameters, but does not alter patient survival.
- Treat underlying cause.

Dyslipidaemia

See Chapter 2.1.

Excessive glucose absorption

Clinical signs

- Weight gain.
- Obesity.
- Elevated lipid levels.

Mechanisms

About 60–80% of the glucose instilled in the peritoneal cavity is absorbed (this is about 50 g with a 4.25% bag, 30 g with a 2.5% bag and 20 g with 1.5% bag). The glucose load contributes to about 20% of the daily energy intake (100–300 g glucose) and while it may be beneficial, it may also contribute to obesity, hyperinsulinaemia and hyperlipidaemia.

Management

- Limit use of PD fluid containing high concentrations of glucose and consider using non-glucose based solutions (e.g. icodextrin).
- Reduce caloric intake.
- Exercise.

Constipation

Mechanisms

Constipation may be less frequent in PD patients than in HD patients, but it may have clinical consequences in PD patients, including catheter malfunction and possibly peritonitis.

Management

- First-line therapy: bulk-forming agents: *Psyllium* husk, *Sterculia*, ispaghula husk.
- Second-line therapy: stool softeners, docusate.
- Third-line agents: osmotic agents, lactulose, sorbitol. If these fail, then use a stimulant (e.g. Coloxyl with senna) and, if required, a glycerine suppository.
- Fourth-line therapy: enema (e.g. Microlax).

Psychological and social problems

Non-compliance

Clinical signs

Chronic fluid overload or inadequate dialysis may be one of the clinical signs of non-compliance. However, detection of non-compliance in the PD patient can sometimes be difficult and may only be identified on a review of inventory of PD solutions at the patient's home. This is one reason for home visits. Non-compliance is probably very common. One study reported that 30% of patients perform fewer than 90%

of their prescribed treatments. Predictors of non-compliance include persistent smoking.

Management

- Encouragement through good education and regular home visits.
- Improved social support.

Patient and carer fatigue

This is an important problem to recognise but can be difficult to overcome. Management includes:

- referral to the renal unit social worker for discussion and advice on coping strategies
- arranging temporary respite care
- sharing responsibilities with more than one carer
- conversion from CAPD to APD
- permanent transfer to a nursing home

Cardiopulmonary compromise

PD produces iatrogenic ascites, which may interfere with the respiratory excursion of the diaphragm and compromise ventilatory function. Forced vital capacity decreases with an increase in IP dialysate volume of more than 1 L. There is also a progressive decrease in LV systolic function. This does not pose a significant problem in patients with normal respiratory function or mild chronic lung disease.

Chapter 4.10
Peritoneal dialysis in children

HIGHLIGHTS

- PD is the most common chronic dialysis therapy worldwide for children.
- The general principles of PD in children are similar to those in adults, except that 'carers' are more likely to be involved and there is greater use of APD.

Aspects of chronic peritoneal dialysis unique to children

Access

- PD access in children is similar to that in adults except that some surgeons advocate removal or trimming of the omentum as this is a frequent cause of catheter obstruction.
- Mechanical complications include inguinal and umbilical hernias and these should be reviewed by a surgeon. Leakage implies that the PD tract has not sealed sufficiently. This may occur if the catheter is used too soon after insertion and in small children and infants who have a thin abdominal wall.

Prescribing and modifying peritoneal dialysis in children

- Initiation of PD: This is best done gradually, beginning with 10 mL/kg per dwell and increasing this by 10 mL/kg over several days as tolerated up to 40 mL/kg.

- Most children now use a cycling machine to provide APD overnight and a 'wet' day.
- Measurement of PD efficacy in children follows the same principles as in the adult.
- As in adults, it is important to establish a dry weight and to review this regularly. The dry weight may increase with growth or be lost with poor nutrition.

Peritonitis

- Peritonitis is the major limiting problem for PD.
- As in adults, prevention is best. This includes aseptic technique, training of all people involved in the management of the PD, and closed line delivery systems used on cycling machines.
- Treatment principles are the same as in adults except that drug doses need modification. In unresponsive infections, particularly due to *Pseudomonas*, the catheter may need to be removed and the child placed on HD, as in adults.

Other medical complications

Nutrition

Children on long-term PD require close monitoring of their nutrition. They have a reduced appetite and increased catabolic demands due to renal failure and because of growth, as well as dietary restrictions. It is important to have a dedicated dietitian to oversee their intake and monitor their growth. Specific issues may include increased or decreased fluid intake, caloric supplements and supplements of vitamins and iron.

Growth

Poor growth is a manifestation of many factors. Treatment includes improving nutrition, correction of acidosis, and monitoring and treatment of bone disease. Occasionally, human growth hormone is of use in children.

Hypertension

HT is less of a problem in children on PD than on HD. Fluid is a major cause, as is renin–angiotensin excess, and management principles are similar to those in adults. Occasionally, prophylactic nephrectomy is required to manage resistant HT. In children with longstanding or severe HT, cardiac monitoring by echocardiography is suggested.

Peritoneal dialysis in children with acute renal failure

PD is frequently used in the acute setting of children with renal failure. Its major advantages are ease of introduction and minimal cardiovascular disturbance. It is, however, less efficient in the removal of toxins and, in this case, HD or CVVHDF is preferred. It is also the preferred modality of delivery of chronic dialysis to small children because of its relative safety with respect to fluid shifts when compared to HD.

Part 5
Acute dialysis

The aim of this part is to address briefly the requirements for acute dialytic therapies, with a particular emphasis on those aspects that are different to chronic dialysis.

Chapter 5.1
Dialysis in acute renal failure

HIGHLIGHTS

- The identification of potentially reversible causes of ARF may obviate the need for dialysis.
- Prescription for acute dialysis is complicated and involves the appropriate choice of dialysis modality, access, dialyser, blood flow rate, UF rate, replacement fluid, anticoagulation, and drug dosing.
- Serious complications should be anticipated to allow optimal management.

Introduction

- The principles underlying the approach to the patient with ARF are given in Protocol 5.1A.
- Certain complications of ARF may require emergency treatment before dialysis can be instituted. See Table 5.1A for a brief outline of the principles underlying their management.
- An early decision is required regarding the need for dialysis and its timing. See Table 5.1B.
- An early decision is required regarding the best modality for acute dialysis (intermittent HD, CVVHDF and rarely PD). See Table 5.1C.
  - CVVHDF allows slow removal of fluid (UF) and is better tolerated in haemodynamically unstable patients and/or those with large fluid requirements.
  - Anticoagulant requirements are less in patients with intermittent HD.
  - PD is indicated in adults only where CVVHDF and intermittent HD are unavailable.
- Hepatitis B, hepatitis C and HIV serology should be confirmed before dialysis, or should be assumed positive until proven otherwise.

Principles of solute and fluid removal in acute renal failure

- Intermittent HD and PD are described in detail in Parts 3 and 4. The principles of fluid and solute removal are similar for ARF.
- With CVVHDF, solute removal is by diffusion (dialysis, CVVHD) and solvent drag of convection (filtration, CVVHF). The simple inclusion of dialysis fluid running on the opposite side of the dialysis membrane converts CVVHF to CVVHDF (See Fig. 5.1). This allows more efficient solute removal. Typical UF rates (the main determinant of solute removal) for CVVHDF are 1–2 L/h.
- Increasingly popular for ARF is a hybrid form of therapy called SLEDD (sustained low-efficiency daily dialysis), in which intermittent dialysis is administered more 'gently', with lower dialysis and blood flow rates for up to 8–12 hours per day.

Table 5.1A Guidelines for the emergency treatment of complications of acute renal failure that may be required prior to dialysis

Complication	Treatment
Hyperkalaemia (K >6.5 mmol/L with ECG changes)	• Cardiac monitor and recheck potassium level in 1–2 hours. • Calcium chloride/gluconate (10 mL of 10% solution over 5 minutes into patent cannula, to avoid extravasation injury) if ECG changes and/or rapid rise, to prevent cardiac toxicity. May need repeating in 30 minutes. • Insulin/dextrose infusion (50 mL 50% dextrose +10 units Actrapid IVI). Give dextrose first and monitor BSL. May need repeating in 2–4 hours. • Sodium bicarbonate if concomitant metabolic acidosis to shift potassium into cells (100 mL 8.4% IV over 30 minutes), if not volume overloaded. May need repeating in 4 hours. • Sometimes, a β_2-agonist (e.g. salbutamol 5 mg nebulised solution) is needed.
Severe volume overload	• Oxygen, positive airways pressure (e.g. CPAP). Intubation. • IV nitrates, starting at 5 μg/min infusion. • High-dose loop diuretic (e.g. frusemide), starting at 100 mg IVI over 30 minutes and, if needed, in 1 hour another bolus of up to 250 mg over 60 minutes, or infusion of 0.5–2.0 mg/min.
Severe hypertension	• Parenteral antihypertensives (e.g. IV hydralazine, 5–10 mg slowly or infusion of 0.2–0.3 mg/min; clonidine, 0.15–0.3 mg over 5 minutes).
Severe metabolic acidosis (e.g. pH < 7.0–7.1)	• Sodium bicarbonate (100 mL 8.4% IV over 30 minutes) if not volume overloaded.
Pericardial tamponade	• Pericardiocentesis
Uraemic seizures	• Diazepam, phenytoin (as for non-uraemic seizures)

PROTOCOL 5.1A

How to initially manage the patient with acute renal failure

- Identify and treat life-threatening complications of ARF, especially hyperkalaemia (see Table 5.1A).
- Identify and treat reversible causes of ARF, such as prerenal azotaemia and lower urinary tract obstruction.
- Identify causes of ARF that require additional urgent treatment, such as vasculitis and rapidly progressive glomerulonephritis.
- Cease all potentially nephrotoxic drugs and adjust doses of renally excreted drugs.
- Differentiate ARF from acute-on-chronic renal failure.
- Decide on the need for and timing of dialysis, and inform the renal physician and renal dialysis unit.
- Choose the most appropriate modality for the first dialysis treatment.

Table 5.1B Guidelines for starting dialysis in acute renal failure

Timing of dialysis	Factors
Dialysis required as soon as possible (within hours)	Fluid overload resistant to conservative measures Hyperkalaemia with ECG changes, especially if rapid rise Severe metabolic acidosis, especially if volume overloaded Some poisonings, especially if ARF (see Chapter 5.2)
Dialysis required soon (by the next day)	In anticipation of and to prevent the above events Severe uraemic complications, including pericarditis and encephalopathy (e.g. altered sensorium, asterixis)
Dialysis required less urgently	In anticipation of and to prevent the above events To keep urea level <30 mmol/L

PROTOCOL 5.1B

How to prepare a patient with acute renal failure for dialysis

- Check hepatitis and HIV serology.
- Assess bleeding risk (coagulation status and platelet count). Consider DDAVP for patients with severe uraemia and/or uraemic bleeding diathesis to reduce the risk of bleeding with catheter placement.
- Place CVC for dialysis access (see Chapter 3.4). Check position of catheter by CXR. Consider femoral approach when there is a risk of bleeding or patient is unable to lie flat.
- Consider phenytoin for patients with a low threshold for seizures.
- Choose the dialysis membrane based on the acuity and severity of renal failure, and medications (see text).
- Reduce the blood flow rate and dialysis times for the initial treatment of patients with severe uraemia.
- Choose the dialysate based on serum biochemistry and comorbidity.
- Choose the anticoagulation regimen based on the bleeding tendency of the patient.

Table 5.1C Criteria for choosing dialysis modality in acute renal failure

Dialysis modality	Criteria
Intermittent HD	Dialysis unit and staff available Need for rapid solute or toxin removal Risk of bleeding Need for frequent movement from ward (e.g. for tests, surgery) Need for prolonged dialytic support
CVVHDF	Intensive care or high-dependency unit Fluid overload Neurological disease Haemodynamic instability Sepsis, ARDS Hypercatabolism, burns
PD	When intermittent HD and CVVHDF are not available

Initiation of dialysis for acute renal failure

Vascular access

Vascular access is discussed in detail in Chapter 3.4. Additional points to be considered with ARF include:

- non-cuffed catheters may be suitable when access is likely to be required for less than 2 weeks
- tunnelled, cuffed 'permcath' catheters should be used when access is likely to be required for longer than 2–4 weeks
- triple-lumen catheters have the advantage of an extra lumen to facilitate blood sampling and drug administration
- cannulation of the femoral (rather than jugular or subclavian) vein is the safest approach for patients with bleeding diathesis, who are unable to tolerate the Trendelenberg position needed for jugular or subclavian cannulation (e.g. because of pulmonary oedema) or due to the unavailability of X-ray facilities to confirm the catheter position, as required for jugular or subclavian catheters

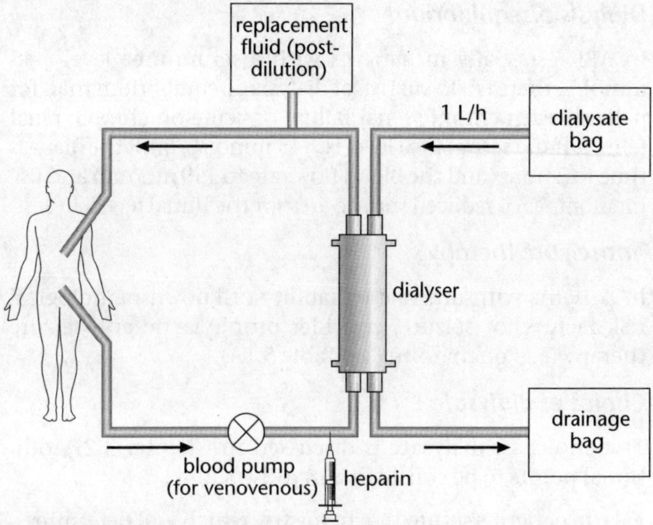

Fig. 5.1 Circuitry for CVVHF and CVVHDF

- Desmopressin acetate (DDAVP, 0.3 µg/kg IVI, to a maximum daily dose of 4 µg/kg) may reduce the risk of bleeding in patients with uraemic bleeding diathesis (see Protocol 2.3B).

Dialysis membranes

Dialysis membranes are discussed in detail in Chapter 3.2. Additional points to be considered with ARF include:

- synthetic biocompatible dialysis membranes may give better outcomes in patients with ARF than unsubstituted or substituted cellulose membranes
- there is a risk of anaphylactoid reactions in patients on ACEIs and in using acrylonitrile-sodium methallyl-sulfonate (AN-69) membranes. (These membranes are used infrequently in Australia.)

Dialysis disequilibrium

In ARF, especially in patients with a serum urea level <45 mmol/L, there is a lower risk of dialysis disequilibrium than for patients with chronic renal failure or acute-on-chronic renal failure. If the serum urea level is >45 mmol/L, limit the dialysis time to 2 hours and the blood flow rate to 150 mL/min and use membranes of reduced surface area for the initial few dialyses.

Antiseizure therapy

In patients with uraemic irritability or known neurological risk factors for seizure, consider prophylactic anti-seizure therapy (e.g. phenytoin) (see Table 5.1A).

Choice of dialysate

The choice of dialysate is discussed in Chapter 3.2. Additional points to be considered in ARF include:

- the patient's serum biochemistry, which will determine choice
- the sodium concentration should not differ from serum levels by more than 15 mmol/L; adequate sodium will facilitate vascular refilling during fluid removal, and prevent hypotension during UF
- the serum bicarbonate and potassium levels should not be corrected too quickly to avoid cardiac arrhythmias and haemodynamic instability
- with severe metabolic acidosis, the potassium level should be reduced carefully by dialysis due to its intracellular shift accompanying the correction of acidosis
- in patients with liver failure or shock, the use of lactate as the dialysate buffer may worsen hyperlactataemia due to its impaired metabolism.
- a dialysate temperature of 36–36.5°C may prevent peripheral vasodilatation and hypotension

See Tables 5.1D and 5.1E for guides to dialysate ion concentrations.

Anticoagulation therapy

Anticoagulation therapy is discussed in Chapter 3.5. Additional points to be considered in ARF include:

Table 5.1D Guide to dialysate ion concentrations[a] for acute renal failure

	Patient (P)	Dialysate (D)[a]
Sodium	125–155	140
	<125, >155	P − D <15
Potassium (K)	<5	3
	>5	2, + Glucose 5[b]
Calcium[c]	<2.4	1.5
	2.4–3.0	1.3
	>3.0	0.9
Magnesium	<0.65	0.5–0.75
	>0.65	0.45 or 0.5

(*a*) All concentrations in mmol/L; see protocols in dialysis unit for amount of additive required to achieve final concentration.
(*b*) If cardiac arrhythmia, use K = 3 mmol/L for first 2 hours, then K = 2 mmol/L. A lower glucose concentration (5 v. 10 mmol/L) may reduce intradialytic potassium movement into cells.
(*c*) Total calcium concentration for patient; ionised calcium concentration for dialysate.

- for patients not at an increased risk of bleeding, use a heparin bolus dose of 500 U followed by 1000 U/h infusion
- for patients with a mildly increased risk of bleeding, use 250 U followed by 250–500 U/h
- to avoid heparin, use intermittent HD, maintain BP with inotropes, flush hourly with normal saline (250–500 mL, depending on amount of clot formed in preceding hour) and consider membranes with possible reduced thrombogenicity (e.g. polyamide)

Continuing dialysis in acute renal failure

- In patients with ARF, more frequent (e.g. daily) or regular (e.g. second daily as opposed to thrice weekly) intermittent HD is required than in patients with chronic renal failure because of hypercatabolism, haemodynamic instability and nutritional needs.
- Hypotension, nephrotoxic agents (e.g. drugs and radiocontrast media) and perhaps bioincompatible

Table 5.1E Ion concentrations in commonly used dialysates[(a)]

	Potassium	Calcium	Magnesium	Glucose
Gambro				
A331	1	1.3	0.2	10
A357	2	1.5	0.35	10
A400	1	0.9	0.2	10
Fresenius				
2886	2	1.5	0.5	5
2883	1	1.3	0.5	5

(*a*) All concentrations in mmol/L. See protocols in dialysis unit for amount of additive required to achieve final concentration higher than these values.

membranes should be avoided to optimise the chances and speed of renal recovery.

- Complications of dialysis are discussed in detail in Chapter 3.7 and are similar for ARF. However, due to the comorbidity and multi-organ involvement of many patients with ARF, the risks are higher (e.g. bleeding, haemodynamic instability, errors of fluid and electrolyte balance, infection, inadequate solute removal).

Drug dosing

- Required adjustments in dosing vary according to the characteristics of the drug and modality of dialysis.
- Drug levels require regular monitoring (e.g. aminoglycosides, vancomycin, digoxin).
- Comprehensive tables for dose adjustment are available and should be consulted for each medication.

References

1. Lennon A-M, Coleman PL, Brady HR. General management of acute renal failure. In: Johnson RJ, Feehally J, eds. Comprehensive clinical nephrology. 2nd edn. London: Mosby, 2003.
2. Aronoff G, Aronoff GR, Berns JS, et al. Drug prescribing in renal failure: dosing guidelines for adults. 4th edn. Philadelphia, PA: American College of Physicians, 1999.
3. Johnson CA, Simmons WD. Dialysis of drugs. 2004 edn. Ann Arbor, MI: Nephrology Pharmacy Associates Inc, 2004.

Chapter 5.2
Extracorporeal therapy for acute poisoning

HIGHLIGHTS

- Historical, clinical and laboratory assessment is necessary to determine the need for extracorporeal decontamination.
- Extracorporeal techniques are ineffective for drugs with large volumes of distribution.
- Haemoperfusion is more effective than haemodialysis for toxins of high MW, protein binding and lipid solubility.

Indications for extracorporeal therapy

Less than 0.05% of acute poisonings are of sufficient severity to warrant the use of extracorporeal decontamination techniques. The principles of initial management of acute poisoning include:

- a thorough exposure history
- a complete physical assessment looking for respiratory, cardiovascular and neurological complications of intoxication
- respiratory and cardiovascular resuscitation
- identification of biochemical (i.e. electrolyte, renal, hepatic, muscle) abnormalities
- identification of poison (e.g. serum anion and osmolal gap, 'toxic screen' of blood and urine, drug levels)
- an assessment of the need for decontamination techniques (e.g. activated charcoal, forced diuresis, extracorporeal therapy)

Determinants of the need for extracorporeal therapy include the amount of poison ingested, the systemic response, the comorbidity and age of the patient and the nature of the poison, balanced against the risks of therapy (Tables 5.2A and 5.2B).

Efficacy

- The relative characteristics that promote extracorporeal drug elimination according to the modality employed are outlined in Table 5.2B.
- The efficacy of extracorporeal therapy is limited when most of the ingested drug is stored outside the extracellular fluid because of high lipid solubility and/or

Table 5.2A When extracorporeal therapy should be considered for poisoning

• When patient is unresponsive to more conservative therapies.
• With clinical evidence of severe toxicity, including coma, haemodynamic instability, severe acidosis, hypothermia, respiratory depression.
• When it will increase total body clearance by ≥30%.
• With impaired endogenous clearance of toxin due to concomitant hepatic or renal dysfunction or known genetic defect of metabolism.
• With increased susceptibility due to intercurrent illness and extremes of age.
• When the perceived benefit outweighs the risk of complications.

Table 5.2B Characteristics of a poison that favour clearance by extracorporeal therapy

• Low MW
• Low protein binding (<30%)
• Small volume of distribution (<1 L/kg)
• Single compartment kinetics
• Low endogenous clearance (<4 mL/min/kg)

tight tissue binding, as is the case with tricyclic antidepressants, digoxin and calcium channel blockers.

- For drugs with prominent tissue stores, these techniques only remove a fraction of the total body load, resulting in 'rebound' of the toxin levels following discontinuation of therapy.

Choice of modality

- Extracorporeal modalities include HD, HF and HP.
- The choice of modality depends on local expertise and availability, and on characteristics of the poison (see Tables 5.2B and 5.2C).
- In addition to accelerating drug clearance, these modalities are useful in the presence of coexisting acute renal failure, fluid overload, metabolic acidosis and electrolyte abnormalities. This is particularly important in preventing seizures and cardiac arrhythmia following tricyclic antidepressant overdose.
- During HD, the toxic compounds are removed by diffusion. This is the most commonly used modality. (See Chapters 4.1 and 5.1 for a technical description.) It is relatively contraindicated in haemodynamically unstable patients.
- HF is commonly delivered by CVVHDF, whereby toxins are removed by diffusion and convection (see Chapter 5.1 for technical description). Although HF is more effective than HD in removing compounds of higher MWs (up to 50000 Daltons, depending on the haemofilter used), it is less efficient for smaller compounds.
- The efficacy of CVVHDF is greatly exceeded by that of HD and HP due to its slower flow rates and lesser solute clearances. However, it is very useful in the treatment of haemodynamically unstable patients.
- During HP, blood circulates through an extracorporeal circuit containing an adsorbent column, such as activated charcoal or polystyrene resin. Adsorbents are incorporated into thin, highly porous membranes that provide a large surface area to bind toxins directly.

Table 5.2C Factors that determine choice of modality of extracorporeal therapy

	Haemodialysis	Haemofiltration	Haemoperfusion
Molecular weight of drug (Daltons)	<500	Depends on filter (<5000–50 000)	Depends on column adsorption
Protein binding >30%	N	± depends on filter	± depends on column
Solubility in water (W) v. lipid (L)	W >>> L	W >> L	W > L
Suitable for haemodynamically unstable patient	±	Y	±
Common examples	Salicylate Lithium Ethanol, methanol Ethylene glycol Diethylene glycol Procainamide[(a)] Disopyramide Selected beta-blockers: atenolol, nadolol, sotalol	Aminoglycosides	Carbamazepine Theophylline Barbiturates Procainamide[(a)] Chloramphenicol Paraquat *Amanita* mushroom

N = not suitable; Y = suitable; ± = equivocal; > = greater than.
(*a*) HP is more effective than HD.

Clearance rates with HP usually exceed those achieved with HD, yet the extraction ratio depends on the ability of the adsorbent to bind the toxic compound.

- PD is much less effective for extracorporeal therapy and is used only where other methods are unavailable, contraindicated or impossible (e.g. in neonates).
- HP can extract drugs that are adsorbed by activated charcoal, with the rate of toxin removal achieved exceeding that achieved with HD when the toxin is highly protein bound, has a high MW or is lipid-soluble. For example, theophylline has an extraction ratio with HD of approximately 50% compared with 99% at the beginning of HP (before the cartridge becomes saturated).

Procedure for haemoperfusion

- The femoral approach is frequently preferred for extracorporeal therapy in view of the likely short duration of treatment, ease of insertion and lower short-term complication rate (see Protocol 3.4A).
- Charcoal or polystyrene resin HP requires specialised cartridges. The most frequently used charcoal cartridge for adults is the Gambro ADSORBA 300C with a priming volume of 260 mL; in children or smaller adults it is the Gambro ADSORBA 150C with a priming volume of 140 mL.
- Protocol 5.2 shows how to initiate and perform charcoal HP. Priming with 5% dextrose reduces the risk of hypoglycaemia by coating with glucose. Heparin requirements are higher than with HD due to the adsorption of heparin to the filter (e.g. approx. 2000 U/h, to keep APTT approx. twice normal). Thus, active haemorrhage or coagulopathy are relative contraindications to HP.
- The complications of charcoal HP are shown in Table 5.2D.
- IV drugs should be administered into the circuit after the filter to prevent adsorption of drugs onto the filter. Drug doses may need to be increased.
- HP is rarely continued for more than 3 hours as the filter becomes saturated with adsorbed toxin.

PROTOCOL 5.2

How to initiate and perform charcoal haemoperfusion

- Temporary vascular access required—femoral approach preferred.
- Read the manufacturer's instructions.
- Set up the circuit as for HD but prime arterial blood lines with 5% dextrose.
- Flush circuit with 2 L of heparinised (2500 U/L) normal saline.
- Initiate HP at a blood flow rate of 50–100 mL/min and gradually increase to 300 mL/min if tolerated.
- Measure the blood glucose level every 30 minutes.
- Measure electrolyte and calcium levels every 60 minutes.
- Measure the platelet count every 60 minutes.
- Monitor the body temperature.
- Look for clotting in the charcoal filter.

Table 5.2D Complications of charcoal haemoperfusion

• Due to adsorption to filter: ○ Hypoglycaemia ○ Thrombocytopenia ○ Reduced drug levels ○ Hypocalcaemia
• Clotting of filter
• Hypothermia (no warmed dialysate, as with HD)
• Leucocytopenia (transient, by up to 10%)
• Rebound toxicity
• Complications of HD and CVVHDF

References

1. Daugirdas JT, Blake PT, Ing TS. Handbook of dialysis. 3rd edn. London: Lippincott Williams & Wilkins, 2000.
2. Rose BD, ed. UpToDate. Version 12.2. Online. Available: <http://www.uptodate.com>.
3. Feehally J. Poisoning and drug overdose. In: Johnson RJ, Feehally J, eds. Comprehensive clinical nephrology. 2nd edn. London: Mosby, 2003.
4. Aronoff G, Aronoff GR, Berns JS, et al. Drug prescribing in renal failure: dosing guidelines for adults. 4th edn. Philadelphia, PA: American College of Physicians, 1999.
5. Haddad LM, Shannon MW, Winchester JF. Clinical management of poisoning and drug overdose. 3rd edn. Philadelphia: WB Saunders, 1998.

Chapter 5.3
Plasmapheresis

HIGHLIGHTS

- Plasmapheresis can rapidly remove pathogenic components from plasma (e.g. antibodies and immune complexes).
- Concurrent immunosuppression to prevent rebound resynthesis of pathogenic antibodies is usually required.
- Serious complications are rare and are more common when using FFP as the replacement fluid versus human albumin.
- Daily medications should be withheld until after plasmapheresis.

Principles

- Apheresis is the process of removing a component of circulating blood using centrifugation, filtration or selective adsorption.
- In plasmapheresis (plasma exchange), circulating plasma containing pathogenic antibodies and immune complexes is removed and exchanged for a replacement fluid, which is usually an equivalent volume of 4.5% human albumin with or without FFP.
- One plasma volume is usually exchanged during each plasmapheresis. The plasma volume of an average adult male is 50 mL/kg body weight. If the predicted plasma volume is 4 L, each 4-L isovolumetric plasmapheresis will reduce preprocedure plasma constituents by two-thirds.
- If there were no extravascular-to-intravascular fluid shift and no new synthesis of antibodies, then three

exchanges would be expected to reduce the baseline plasma constituents by 95%.

- IgG is distributed in equal amounts in both the intra- and extravascular spaces. If new antibody production can be suppressed, then six plasmaphereses would be required to reduce total body IgG by more than 95%.
- Because IgM and large circulating immune complexes are mainly limited to the intravascular space, they are far more effectively removed by plasmapheresis.

Indications

Tables 5.3A and 5.3B show the established and possible roles for plasma exchange.

Regimens

Protocol 5.3 shows how to prescribe plasma exchange in renal disease.

Complications

The overall complication rate is around 1.4% when using albumin as replacement fluid, compared to 20% when using FFP. Most of these complications are not serious.

Bleeding

- Due to the depletion of coagulation factors, the pro-thrombin time is increased by 30% and partial thromboplastin time by 100% after a single plasma volume exchange. If there is a risk of serious bleeding, 5–8 mL/kg of FFP should be used as part of the exchange volume.

Table 5.3A Established role for plasma exchange

Condition	Role
Goodpasture's disease	Pulmonary haemorrhage and non-dialysis-dependent oliguric renal failure are strong indications
ANCA-positive vasculitis	Pulmonary haemorrhage and dialysis-dependent renal failure are strong indications
HUS/TTP	May allow for greater replenishment of von Willebrand factor cleaving protease as well as removal of its inhibitor antibodies
Cryoglobulinaemia	Very effective for rapid control of symptoms. Indicated when there is evidence of vasculitis involving nerves or kidneys. Treatment for underlying lymphoma or hepatitis C is still required.
Waldenstrom macroglobulinaemia	Very effective for hyperviscosity syndrome
Myasthenia gravis Guillain-Barré syndrome Chronic inflammatory demyelinating polyneuropathy Stiff person syndrome	
ITP Post-transfusion purpura and coagulopathy due to coagulation factor inhibitors	

Table 5.3B Possible role for plasma exchange

Condition	Role
Highly sensitised transplant recipients Antibody-mediated allograft rejection	Remove preformed anti-HLA antibodies
Myeloma with cast nephropathy Crescentic IgA nephropathy Recurrent focal segmental glomerulosclerosis	 Transient reduction in proteinuria
Catastrophic antiphospholipid antibody syndrome Crescentic renal lupus or cerebral lupus	 No significant benefit for treating moderately severe or mild renal lupus
Pemphigus vulgaris HELLP syndrome Anti-Ro antibodies in pregnancy	

PROTOCOL 5.3

How to prescribe plasma exchange in renal disease

Indication	When to start	Duration	Replacement fluid[a]	Concurrent immunosuppression
Goodpasture's (anti-GBM) disease	If the patient is not already dialysis dependent, or if there is pulmonary haemorrhage even with dialysis-dependent renal failure	Daily exchange until anti-GBM antibodies <20%, requires at least 7 exchanges	Normal saline with 5% albumin, 4 mmol KCl/L if not dialysis dependent, 4.6 mmol calcium gluconate/L	Prednisone and cyclophosphamide
ANCA-positive small-vessel vasculitis	Pulmonary haemorrhage or dialysis dependent RPGN	Daily to second daily to achieve at least 5 exchanges over 7 days, more depending on clinical response	As above	As above
Cryoglobulinaemia (type II)	MPGN or mononeuritis multiplex	Daily exchange until there is a satisfactory clinical response	As above	Interferon/ribavirin for HCV, prednisone + cyclophosphamide if severe vasculitis

Indication	When to start	Duration	Replacement fluid[a]	Concurrent immunosuppression
HUS/TTP	When diagnosis is made	Daily until platelet count is $>80 \times 10^9$/L and there is no more fragmentation on blood film	FFP	Nil

One exchange volume = 50 mL/kg body weight.
(*a*) If there is any bleeding or risk of bleeding (e.g. renal biopsy within 48 hours, thrombocytopenia, coagulopathy), then FFP 5–8 mL/kg should be used as part of the replacement fluid.

Anaphylactoid reactions

- These can be serious. There is an increased risk when using FFP rather than albumin. Premedication with 100 mg IV hydrocortisone and an H_1-receptor antihistamine can help.

Hypocalcaemia

- Citrate used in the extracorporeal circuit as an anticoagulant binds to free calcium, which can cause symptomatic hypocalcaemia. Additionally, the metabolism of citrate to bicarbonate can cause metabolic alkalosis in patients with renal impairment.

Hypogammaglobulinaemia

- This can predispose to bacterial infections, especially sinusitis and pneumonia. Patients should not be given live vaccines (see Chapter 2.6).

Removal of protein-bound medications

- Plasmapheresis should be withheld for at least 24 hours following bolus IV cyclophosphamide to ensure drug effect. Daily medications, especially those that have high protein binding, should be withheld until after plasmapheresis.

Vascular access problems

- Infection, bleeding and inadequate access.

Test results

- False negative and positive results can occur for tests relying on measuring serum protein levels (e.g. viral serology, autoimmune serology, hormone levels, etc.), which may be removed by plasmapheresis.

References

1. Clark W, Rock G, Buskard N, et al. Therapeutic plasma exchange: an update from the Canadian Apheresis Group. Ann Intern Med 1999; 131: 453–62.
2. Rich RR, ed. Clinical immunology. Principles and practices. London: Mosby, 1996: 2068–82.
3. Levy J, Pusey C. Plasma exchange. In: Johnson RJ, Feehally J, eds. Comprehensive clinical nephrology. 2nd edn. London: Mosby, 2003.

Index